SpringerBriefs in Geography

SpringerBriefs in Geography presents concise summaries of cutting-edge research and practical applications across the fields of physical, environmental and human geography. It publishes compact refereed monographs under the editorial supervision of an international advisory board with the aim to publish 8 to 12 weeks after acceptance. Volumes are compact, 50 to 125 pages, with a clear focus. The series covers a range of content from professional to academic such as: timely reports of state-of-the art analytical techniques, bridges between new research results, snapshots of hot and/or emerging topics, elaborated thesis, literature reviews, and in-depth case studies.

The scope of the series spans the entire field of geography, with a view to significantly advance research. The character of the series is international and multidisciplinary and will include research areas such as: GIS/cartography, remote sensing, geographical education, geospatial analysis, techniques and modeling, landscape/regional and urban planning, economic geography, housing and the built environment, and quantitative geography. Volumes in this series may analyze past, present and/or future trends, as well as their determinants and consequences. Both solicited and unsolicited manuscripts are considered for publication in this series.

SpringerBriefs in Geography will be of interest to a wide range of individuals with interests in physical, environmental and human geography as well as for researchers from allied disciplines.

More information about this series at http://www.springer.com/series/10050

Blal Adem Esmail · Davide Geneletti

Ecosystem Services for Urban Water Security

Concepts and Applications in Sub-Saharan Africa

 Springer

Blal Adem Esmail
Department of Civil, Environmental and
Mechanical Engineering
University of Trento
Trento, Italy

Division of Urban and Regional Studies
Department of Urban Planning and
Environment, Royal Institute of
Technology
Stockholm, Sweden

Davide Geneletti
Department of Civil, Environmental and
Mechanical Engineering
University of Trento
Trento, Italy

ISSN 2211-4165 ISSN 2211-4173 (electronic)
SpringerBriefs in Geography
ISBN 978-3-030-45665-8 ISBN 978-3-030-45666-5 (eBook)
https://doi.org/10.1007/978-3-030-45666-5

This Springer imprint is published by the registered company Springer Nature Switzerland AG
The registered company address is: Gewerbestrasse 11, 6330 Cham, Switzerland

Foreword 1

Cities worldwide are facing increasing water pressures as population growth, land degradation and climate changes increase stresses on water supply systems, intensify risks of intense floods and exacerbate periodic droughts. Leaders are increasingly turning to nature-based solutions—actions like investing in sustainable land use practices, improving vegetation cover and protecting existing natural landscapes and floodplains—as part of an integrated strategy to secure water for their people. In recent years, a plethora of tools and approaches have appeared on the scene to help policymakers from national to local scales to decide on the best ways to use nature to meet their water needs.

But, technical tools that help demonstrate the links between land management and water are not enough; there is an urgent need to better understand how, where and in what configuration sustainable land use can contribute to water security in increasingly complex water supply and delivery systems, and what the impacts on local livelihoods will be from different management choices. Few studies have taken such a comprehensive look at the water system—and the role of nature within it—as the one presented here.

Further, despite the availability of many technical tools and models of natural systems and hydrology, leaders and technical staff still lack practical examples of how and where such analyses have been done in their region. This book presents a practical case study for how an integrated evaluation of sustainable land management as part of a city's broader water security strategy can be applied in the context of a rapidly developing city in Sub-Saharan Africa. Such case studies help to build the critical base of knowledge necessary to mainstream the use of technical tools in the day-to-day operations of the world's water supply systems.

Mainstreaming sustainable land management into water security strategies for developing countries is both a huge challenge and a grand opportunity. The challenge is particularly acute in data-poor regions, such as in Sub-Saharan Africa, where a lack of fine-scale data on both ecosystem and the ways that people depend on them makes applying technical assessments challenging. The role of stakeholders and local knowledge in such cases cannot be overstated, to fill the gaps in our knowledge—not just of the working of natural systems, but of appropriate local land management technologies, and the impacts that watershed management programs can have on livelihoods, social interactions, power dynamics and gender roles.

Water security for growing cities is not a problem unique to Africa, nor to the developing world. But by sharing experiences, techniques and lessons learned, we can work together towards making the vision of water systems that sustain both nature and people a reality.

Adrian L. Vogl, Ph.D.
Lead Scientist, The Natural Capital Project (NatCap)
Securing Freshwater Initiative, Stanford University (USA)
Stanford, USA

Foreword 2

The UN World Water Development Report 2018 acknowledged that future investments in nature-based solutions could benefit with mainstreaming of natural infrastructure into planning. Hard engineering strategies to build water security in urban areas have been implemented around the world. These solutions are expensive and can be intrusive to the natural environment, thereby presenting nature-based solutions inspired by the ecosystem services concept—as an alternative. The ecosystem services approach also provides a foundational basis for many *"fit to purpose"* and *"cost-effective"* nature-based solutions. Still, projects and programs inspired by ecosystem services that intent to improve water availability, help enhance water quality and generally maintain our natural capital should be discussed along with conventional planning measure to manage water. In this book, the authors explore and review what conceptual framings and practical options are available to apply ecosystem services approach for water security, and how it could be valid for urban water management in the African region.

For many cities and communities, the way their water is managed will help them executing vulnerability and achieve socio-ecological resilience. Applying an ecosystem services perspective to planning can provide cities' decision-makers, and development partners a new and practical guide for developing effective policies, design and investment plans to address the water security agenda. The UN-Water 2013 Report that interlinks water security into a conceptual framework and captures the complexity inherent to sustainable urban management is gaining global attention. Adopting a water security vision comprehends the need to manage water resources alongside the political, economic and social discourse—especially in case of people and populations in vulnerable contexts. This book reflects well on that argument.

Indeed, planning for ecosystem services in the urban context is gaining momentum as more cities and communities realize that hard engineering solutions are costly and challenging to develop, operate and maintain. States and planners are considering adopting ecosystem services knowledge and have already embarked on similar endeavours—as elucidated by the Asmara and the Toker Watershed case study analysed in this book. The Food and Agriculture Organization of the United

Nations published *Forests and Sustainable Cities* (Unasylva 250), in 2018, highlighting the need for mapping opportunities to deploy smart, green, cost-effective water management policies in towns and cities. This book is a timely and an excellent attempt to link the water security agenda with the ecosystem services framework for urban water security planning and builds well on the conceptual density and existing scholarship, thematically and empirically. It promises to enhance the understanding of water needs in the vulnerable region of Sub-Saharan Africa.

Nidhi Nagabhatla, Ph.D.
Programme Officer & Capacity Building Coordinator
United Nations University—Institute for Water
Environment and Health (UNU-INWEH), Canada
and Adjunct Professor—McMaster University, Canada

Acknowledgements

Research for this book has been conducted as part of a Ph.D. study at the University of Trento, in the Department of Civil, Environmental and Mechanical Engineering. Chapters 3 and 5 draw from Adem Esmail & Geneletti (2017), *Design and impact assessment of watershed investments: An approach based on ecosystem services and boundary work*, published in Environmental Impact Assessment Review (62, 1–13). We are grateful to Dr. Chiara Cortinovis for her helpful suggestions.

Contents

About the Authors

Blal Adem Esmail, Ph.D., Research Scientist at KTH Royal Institute of Technology, Department of Urban Planning and Environment. Formerly, a Postdoc fellow at the University of Trento, Department of Civil, Environmental and Mechanical Engineering, visiting scholar at Leibniz University Hannover, and teaching assistant in Water Engineering.

Davide Geneletti, Ph.D., Associate Professor at the University of Trento, leader of the Planning for Ecosystem Services research group (www.planningfores.com), former research fellow at Harvard University's Sustainability Science Program, and visiting scholar at Stanford University.

List of Figures

List of Tables

Chapter 1
Introduction

Abstract This chapter sets the context of the book by providing a brief account of the challenges and opportunities of urbanization in Sub-Saharan Africa, with a focus on the urban water sector. Watershed investments are here emphasized as a promising opportunity to effect large-scale transformative change promoting human wellbeing while conserving life-supporting ecosystems. The chapter concludes by illustrating the three specific objectives of the book.

Keywords Urbanization · Ecosystem services · Watershed investments · Ecosystem-based response · Integrated urban water management · Water security · Sub-Saharan Africa

1.1 Urbanization, Ecosystem Services and Water Security

In an era of rapid urbanization, cities offer an indispensable perspective to tackle the quest for sustainable development that aims at achieving human wellbeing, while preserving the ecosystems that sustain life on Earth (Kates et al., 2001, William and Levin 2009). Today, cities occupy less than 2% of the global territory, but represent 70% of the economy and 60% of energy consumption, among others. According to the UN World Urbanization Prospect, globally, more people live in urban areas than in rural areas, with 55% of the world's population residing in urban areas in 2018, and 68% of the world's population projected to be urban by 2050 (UN-DESA 2019). Interestingly, almost 90% of the additional 2.5 billion urbanites expected by 2050 will be concentrated in Africa and Asia. The fastest growing urban agglomerations are medium-sized cities and cities with less than 1 million inhabitants (Fig. 1.1).

Challenges of water security in cities, including freshwater scarcity and flooding, represent major global socio-ecological problems of the twenty-first century. Water-related hazards, and generally natural hazards, are estimated to have resulted in billions of dollars in damage and heavy loss of human lives, over the past decades, affecting the sustainability of urban areas and wellbeing of their inhabitants (UNISDR 2015). Hard engineering initiatives and strategies to achieve urban water security have been implemented around the world. However, water security entails significant complexities and uncertainties that no longer can be addressed effectively with traditional

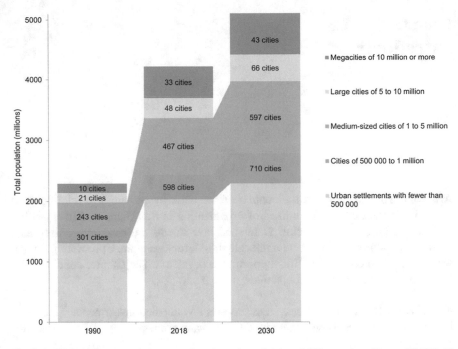

Fig. 1.1 Global urban population growth and number of cities of different sizes (*Source* UN World Urbanization Prospect 2018)

approaches. There is instead a need to shift from current planning and management paradigms of "predictions and control" to one of "adaptation and innovation" (e.g. Pahl-Wostl et al. 2011).

Human wellbeing in cities is substantially linked to goods and services provided by ecosystems (Geneletti et al. 2020). This is well the case of the water sector, in which the role of ecosystem is increasingly recognized alongside that of infrastructures and institutional arrangements (e.g. Adem Esmail et al. 2016). In general, the concept of ecosystem services is emerging as an effective tool to support policy- and decision-making that aims at promoting sustainability, from raising stakeholders' awareness to shaping decisions (e.g. Ruckelshaus et al. 2015; Burkhard et al. 2018). In particular, incorporating ecosystem services knowledge to generate actions and produce outcomes, supporting new policies that explicitly consider effects on ecosystem services is believed to make significant impact towards sustainability, promoting human wellbeing while conserving ecosystems (Cortinovis and Geneletti 2018; Haase et al. 2014; Spyra et al. 2018).

Watershed investments to secure water for cities represent a promising opportunity to effect large-scale transformative changes for sustainable development (Guerry et al., 2015: Kroeger et al. 2019; McDonald and Shemie 2014). They consist of governance and financial mechanisms that secure clean water for downstream users, mainly cities, and operate by engaging upstream communities (Hamel et al. 2020;

Higgins and Zimmerling 2013). They target a wide range of activities—from changes in land use and terracing to education and community outreach—to enhance selected ecosystem services such as erosion control and nutrient retention, while conserving nature and biodiversity. Watershed investments can be an effective way of implementing adaptive management for urban water security. To this end, their design and implementation require adequate operational approaches that duly address the concerns of diverse stakeholders, accounting for their roles and relative influence and power, and consider relevant contextual and contingent factors.

Indeed, the challenges of urban water security are particularly arduous, requiring urgent responses, in the Sub-Saharan Africa region. Over the past decades, unlike the rest of the world, this region has not experienced a steady increase of the level of wellbeing as for, example, measured in terms of Gross National Product. In addition, several studies consistently predicted that the region would face water crises compounded by a significant growth of its urban population (Douglas 2018; Giugni et al. 2015). On the other hand, many of the cities that will host an additional 2.5 billion urbanites expected by 2050, mostly in Africa and Asia, are yet to be built, and this offers a unique window of opportunity to explore novel planning and management paradigms to effectively address the many challenges they will face, including those related to urban water security.

1.2 Aim and Structure of This Book

This book addresses the challenges of urban water security, and specifically of implementing adaptive management in the water sector in Sub-Saharan Africa. To this end, it pursues three specific objectives: (i) developing a conceptual framework of the urban water sector from an ecosystem services perspective, highlighting the specificities of the Sub-Saharan context; (ii) developing an operational approach for designing and assessing impact of watershed investments, based on ecosystem services modelling and boundary work; (iii) testing the proposed approach through a case study in Eritrea.

The book explores and interlinks concepts, such as ecosystem services, boundary work and learning organizations. The concept of ecosystem services provides a holistic approach for framing socio-ecological issues and for integrating different types of data (e.g. biophysical and socio-economic), to identify possible solutions (Geneletti et al. 2020). Boundary work, a subset of the bridging organization literature, is defined as a set of activities put in place by any organization or individual that seeks to mediate between knowledge and action (Cash et al. 2003). It consists of any effort to manage the tension that arises at the interface between stakeholders that have differing views on what represents relevant knowledge (Clark et al. 2016). Finally, a learning organization is an organization that is skilled at creating and acquiring knowledge and modifying its behaviour to reflect new insights (Cowling et al. 2008). Together with the related concept of institutional capacity, it helps to frame the role of social actors, such as water utilities, in structuring their choice of action within a society to pursue their mission (Kayaga et al. 2013).

The book is structured into five parts. Chapter 2 provides an overview of the urban water sector, including both infrastructures and institutions, from an ecosystem services perspective, with some examples of real-life projects of urban water infrastructures in Sub-Saharan Africa. It thus addresses the first specific objective of this book. Chapter 3, related to the second specific objective, focuses on watershed investments for securing water for cities. It starts with a brief account of the application of ecosystem services for decision-making and a theoretical background of boundary work; hence, it develops an operational approach for designing and assessing impact of watershed investments to secure urban water security. The approach combines spatially explicit ecosystem services modelling with insights of boundary work. The two chapters that follow address the third specific objective of this book. Chapter 4 introduces a case study of the urban water sector in Eritrea to demonstrate the application of the approach presented in Chap. 3. The case is about a medium-sized city, Asmara, and its main water supply, the Toker Watershed. The most relevant socio-ecological challenges and opportunities related to urban water security are illustrated, highlighting the crucial role of the Asmara Water Supply Department. In particular, the latter is analysed through a novel approach for conceptualizing water utilities as learning organization and assessing their institutional capacity. Chapter 5 presents an application of the approach for designing and assessing impact of watershed investments, developed in Chap. 3, to the Asmara and Toker Watershed case study. Assuming urban water security and rural poverty alleviation as two objectives for watershed investment, the case study explores all the steps of the proposed approach. The results include spatially explicit data that allow quantitatively assessing the performance of different watershed investment scenarios in terms of changes in a selected ecosystem services, answering to important planning and management questions. The application also highlights the challenges of addressing stakeholders' concerns through relevant boundary work strategies. The book concludes with Chap. 6 summarizing the main messages, as well as discussing the challenges for future research and practice to contribute to achieving urban water security and generally to implementing adaptive management in the urban water sector.

References

Adem Esmail B, Geneletti D (2017) Design and impact assessment of watershed investments: an approach based on ecosystem services and boundary work. Environ Impact Assess Rev 62:1–13. https://doi.org/10.1016/j.eiar.2016.08.001

Adem Esmail B, (2016) Ecosystem services for watershed management and planning. (PhD Thesis). University of Trento, Italy.

Adem Esmail B, Suleiman L. Analysing evidence of Sustainable Urban Water Management Systems: a review through the lenses of sociotechnical transitions. Sustainability.

Burkhard B, Maes J, Potschin-Young M, Santos-Martín F, Geneletti D, Stoev P, Kopperoinen L, Adamescu C, Adem Esmail B, Arany I, Arnell A, Balzan M, Barton DN, van Beukering P, Bicking S, Borges P, Borisova B, Braat L, M Brander L, Bratanova-Doncheva S, Broekx S, Brown C, Cazacu C, Crossman N, Czúcz B, Daněk J, Groot R. de Depellegrin D, Dimopoulos P, Elvinger N, Erhard M, Fagerholm N, Frélichová J, Grêt-Regamey A, Grudova M, Haines-Young R, Inghe O, Kallay T, Kirin T, Klug H, Kokkoris I, Konovska I, Kruse M, Kuzmova I, Lange M, Liekens I, Lotan A, Lowicki D, Luque S, Marta-Pedroso C, Mizgajski A, Mononen L, Mulder S, Müller F, Nedkov S, Nikolova M, Östergård H, Penev L, Pereira P, Pitkänen K, Plieninger T, Rabe S-E, Reichel S, Roche P, Rusch G, Ruskule A, Sapundzhieva A, Sepp K, Sieber I, Šmid Hribar M, Stašová S, Steinhoff-Knopp B, Stępniewska M, Teller A, Vackar D, van Weelden M, Veidemane K, Vejre H, Vihervaara P, Viinikka A, Villoslada M, Weibel B, Zulian G (2018) Mapping and assessing ecosystem services in the EU—Lessons learned from the ESMERALDA approach of integration. One Ecosyst 3:e29153. https://doi.org/10.3897/oneeco.3.e29153

Cash DW, Clark WC, Alcock F, Dickson NM, Eckley N, Guston DH, Jager J, Mitchell RB (2003) Knowledge systems for sustainable development. Proc Natl Acad Sci 100:8086–8091. https://doi.org/10.1073/pnas.1231332100

Clark WC, Tomich TP, van Noordwijk M, Guston D, Catacutan D, Dickson NM, McNie E (2016) Boundary work for sustainable development: natural resource management at the Consultative Group on International Agricultural Research (CGIAR). Proc Natl Acad Sci 113:4615–4622. https://doi.org/10.1073/pnas.0900231108

Cortinovis C, Geneletti D (2018) Mapping and assessing ecosystem services to support urban planning: a case study on brownfield regeneration in Trento. Italy One Ecosyst 3:e25477. https://doi.org/10.3897/oneeco.3.e25477

Cowling RM, Egoh B, Knight AT, O'Farrell PJ, Reyers B, Rouget M, Roux DJ, Welz A, Wilhelm-Rechman A (2008) An operational model for mainstreaming ecosystem services for implementation. Proc Natl Acad Sci 105:9483–9488. https://doi.org/10.1073/pnas.0706559105

Douglas I (2018) The challenge of urban poverty for the use of green infrastructure on floodplains and wetlands to reduce flood impacts in intertropical Africa. Landsc Urban Plan 180:262–272. https://doi.org/10.1016/j.landurbplan.2016.09.025

Geneletti D, Cortinovis C, Zardo L, Esmail BA (2020) Planning for ecosystem services in cities. SpringerBriefs in Environmental Science. Springer International Publishing, Cham. https://doi.org/10.1007/978-3-030-20024-4

Guerry AD, Polasky S, Lubchenco J, Chaplin-Kramer R, Daily GC, Griffin R, Ruckelshaus M, Bateman IJ, Duraiappah A, Elmqvist T, Feldman MW, Folke C, Hoekstra J, Kareiva PM, Keeler BL, Li S, McKenzie E, Ouyang Z, Reyers B, Ricketts TH, Rockström J, Tallis H, Vira B, (2015) Natural capital and ecosystem services informing decisions: From promise to practice. Proc. Natl. Acad. Sci. 112, 201503751. https://doi.org/10.1073/pnas.1503751112

Giugni M, Simonis I, Bucchignani E, Capuano P, De Paola F, Engelbrecht F, Mercogliano P, Topa ME (2015) The Impacts of climate change on African cities, pp 37–75. https://doi.org/10.1007/978-3-319-03982-4_2

Haase D, Frantzeskaki N, Elmqvist T (2014) Ecosystem services in urban landscapes: practical applications and governance implications. Ambio 43:407–412. https://doi.org/10.1007/s13280-014-0503-1

Hamel P, Bremer LL, Ponette-González AG, Acosta E, Fisher JRB, Steele B, Cavassani AT, Klemz C, Blainski E, Brauman KA (2020) The value of hydrologic information for watershed management programs: the case of Camboriú. Brazil Sci. Total Environ 135871. https://doi.org/10.1016/j.scitotenv.2019.135871

Higgins J, Zimmerling A (2013) A primer for monitoring water funds. Global Freshwater Program, The Nature Conservancy, Arlington, VA

Kates RW, Clark WC, Corell R, Hall JM, Jaeger CC, Lowel, McCarthy JJ, Schellnhuber HJ, Bolin B, Dickson NM, Faucheux S, Gallopin GC, Grübler A, Huntley B, Jäger J, Jodha NS, Kasperson RE, Mabogunje A, Matson P, Mooney H, Moore III B, O'Riordan T, Svedin U (2001) Environment and development: Sustainability science. Science (80). 292, 641–642. https://doi.org/10.1126/science.1059386

Kayaga S, Mugabi J, Kingdom W (2013) Evaluating the institutional sustainability of an urban water utility: a conceptual framework and research directions. Util Policy 27:15–27. https://doi.org/10.1016/j.jup.2013.08.001

Kroeger T, Klemz C, Boucher T, Fisher JRB, Acosta E, Cavassani AT, Dennedy-Frank PJ, Garbossa L, Blainski E, Santos RC, Giberti S, Petry P, Shemie D, Dacol K, (2019) Returns on investment in watershed conservation: Application of a best practices analytical framework to the Rio Camboriú Water Producer program, Santa Catarina, Brazil. Sci. Total Environ. 657, 1368–1381. https://doi.org/10.1016/j.scitotenv.2018.12.116

McDonald RI, Shemie D, (2014) Urban Water Blueprint: Mapping conservation solutions to the global water challenge. TNC, Washington, DC.

Pahl-Wostl C, Jeffrey P, Isendahl N, Brugnach M (2011) Maturing the new water management paradigm: progressing from aspiration to practice. Water Resour Manag 25:837–856. https://doi.org/10.1007/s11269-010-9729-2

Ruckelshaus M, McKenzie E, Tallis H, Guerry A, Daily GC, Kareiva P, Polasky S, Ricketts T, Bhagabati N, Wood SA, Bernhardt JR (2015) Notes from the field: lessons learned from using ecosystem service approaches to inform real-world decisions. Ecol Econ 115:11–21. https://doi.org/10.1016/j.ecolecon.2013.07.009

Spyra M, Kleemann J, Cetin NI, Vázquez Navarrete CJ, Albert C, Palacios-Agundez I, Ametzaga-Arregi I, La Rosa D, Rozas-Vásquez D, Adem Esmail B, Picchi P, Geneletti D, König HJ, Koo H, Kopperoinen L, Fürst C (2018) The ecosystem services concept: a new Esperanto to facilitate participatory planning processes? Landsc Ecol 6. https://doi.org/10.1007/s10980-018-0745-6

UN-DESA (2019) World urbanization prospects 2018: highlights. Webpage. United Nations

UN Water, (2018) Nature-Based Solutions for Water. United Nations Educational, Scientific and Cultural Organization (UNESCO).

UNISDR (2015) Making development sustainable: the future of disaster risk management. In: Global assessment report on disaster risk reduction. United Nations Office for Disaster Risk Reduction (UNISDR), Geneva, Switzerland

William CC, Levin SA (2009) Toward a science of sustainability. 33:172

Chapter 2
Linking Ecosystem Services to Urban Water Infrastructures and Institutions

Abstract This chapter provides an overview of the urban water sector, including both infrastructures and institutions, from an ecosystem services perspective. Hence, it proposes a conceptual framework intended to highlights the role of urban water sector in (i) linking ecosystem services supply and benefitting areas, (ii) bridging spatial scales ranging from the watershed to the household level and (iii) adopting ecosystem-based responses to water vulnerability. An example of application of the framework is shown using real-life projects of urban water infrastructures in Sub-Saharan Africa. The framework sets a useful background for further analysis of the urban water sector, as presented following chapter focusing on the watershed scale.

Keywords Ecosystem-based response · Integrated urban water management · SEEA-water framework · Urbanization · Water security · Sub-Saharan Africa

2.1 A Conceptual Framework

This chapter proposes a conceptual framework of the urban water sector from an ecosystem services perspective. By urban water sector, we refer to urban water infrastructures and to the institutions that operate and manage them, water utilities in the first place. The proposed framework builds on established approaches and concepts from the urban water sector, such as the Integrated Urban Water Management (Bahri 2012; Cowie and Borrett 2005) and System of Economic and Environmental Accounting for Water (UN-DESA 2011). This is a key strategy to facilitate its application and mainstreaming in real-world contexts.

The proposed framework is shown in Fig. 2.1. In essence, the framework is a synthesis of some of the most significant aspects characterizing the exchange of water between watersheds and cities, and within the city, based on a review of the literature. In particular, the framework is intended to highlights the role of urban water sector in (i) linking ecosystem services supply and benefitting areas, (ii) bridging spatial scales ranging from the watershed to the household level and (iii) adopting ecosystem-based responses to water vulnerability. The framework is structured into four parts, described in the next sections: urban water infrastructures (section 2.1.1),

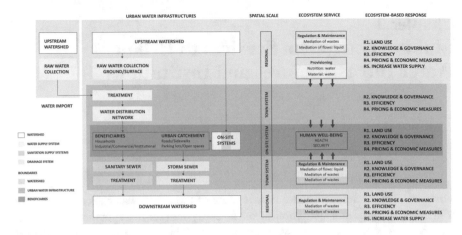

Fig. 2.1 Conceptual framework of the urban water sector from an ecosystem services perspective. Thin black arrows represent the flow of freshwater and wastewater; thicker arrows represent the flow of ecosystem services. (Modified after Adem Esmail 2016)

spatial and temporal scale (section 2.1.2), ecosystem services (section 2.1.3) and (d) ecosystem-based responses (section 2.1.4).

2.1.1   Urban Water Infrastructures

The first part of the framework identifies the urban water infrastructures—water supply, sanitation and drainage systems—consisting of engineered and non-engineered structures, equipment and facilities that are needed to deliver water services for economic production and household use (World Bank 1994). Such infrastructures allow water to flow from the watersheds to the urban beneficiaries (i.e. water supply system) and back to the watersheds (i.e. sanitation and drainage systems). As highlighted in Fig. 2.1, an important distinction is between the so-called "Town systems" and "On-site Systems" or facilities (Choguill 1996, 1999). The former consist of centralized infrastructures, generally built and managed by municipalities to serve the central areas of cities and areas where high-income residences are located. The latter include all the "creative" means by which the poor in underserved areas meet their basic needs related to water supply, sanitations and hygiene; they include pit latrines, septic tanks and drinking-water wells. This dichotomy is in fact characteristic of many cities in the developing world, where a formal and informal sector coexist (WHO and UN-HABITAT 2010).

In this regard, two distinct trends appear to be characterizing the urban water sector. A first trend promoting "progressive improvement" according to which on-site systems can be planned and implemented so that, with time, they will meet desirable standards and eventually become integral part of the town system (Choguill 1996,

1999). With the right mix of policies (e.g. land tenure and infrastructure owner-ship) and proper technical support, communities could in fact self-build their own infrastructures as in the case of the Orangi District of Karachi, in Pakistan, where an unauthorized, low-income community of about 800,000 inhabitants has successfully developed and built its own sewer system (Bahri 2012). Perhaps, this also may be the only viable solution for most cities in the developing world.

The second trend consists in promoting more decentralized solutions, questioning the sustainability of traditional town systems (Larsen and Gujer 1997). The latter, intended as providing potable water and flushing toilets in every household, are in fact, increasingly criticized in several respects: environmental (e.g. resource protec-tion), social (e.g. security of supply) and financial (e.g. cost recovery and afford-ability) (e.g. Larsen and Gujer 1997; Lieberherr and Truffer 2015; Suleiman et al. 2019). Among others, the second trend calls to leapfrog a typical water development pattern in cities, which starts with (a) exhaustion of local water resources followed by (b) water import from adjacent watersheds, hence (c) introduction of water con-servation measures and finally (d) adoption of local solutions, such as storm and rainwater harvesting or seawater desalination (e.g. Richter et al. 2013). Therefore, in Fig. 2.1, "Town Systems" and "On-site system" represent the underlying two trends of "Progressive improvement" and "Decentralization", respectively.

2.1.2 Spatial and Temporal Scales

The second part of the framework represents the spatial scales covered by the urban water sector: from the regional level (e.g. watersheds) to that of the city (e.g. town sys-tems) and finally the household level (e.g. on-site systems). In Fig. 2.1, three shades of grey boundaries define the extents of the watersheds (light grey), urban water infrastructures (grey) and beneficiaries (darker grey). Less noticeably represented are the underlying implications relating to the temporal scales, which span from annual/seasonal variations (e.g. rainfall patterns at the watershed level) to instanta-neous, individual demands for tap water by users (e.g. coping with so-called "toilet-peak" at the household level). For instance, the former determine the availability of water resources and are crucial for planning purposes, while the latter are related to the end-users' demand and perception and are important for management goals. Generally, both the spatial and temporal scales can have significant and complex implications in terms of management, planning and policy-making. Therefore, in the proposed framework, they are considered by highlighting different boundaries (i.e. watershed, urban water infrastructure, beneficiaries' boundary), distinguishing between upstream and downstream areas, including both town system and on-site systems, and specifying the direction of the flow of ecosystem services, among others.

2.1.3   Ecosystem Services

The third part of the framework shows the main ecosystem services intercepted by the different components of the urban water infrastructure, specifying the constituents of human wellbeing to which they contribute. As reported in Table 2.1, these are provisioning and regulating ecosystem services that contribute mainly in terms of security, health and livelihood of urban dwellers and activities. Indeed, several other ecosystem services contribute to human wellbeing in cities; however, the focus here is on those directly intercepted by urban water infrastructures (e.g. for sanitation systems, ecosystem services could be identified considering the components of the systems as reported in Tilley et al. 2014). Last, several classifications of ecosystem services exist, including the ones proposed by (Boyd and Banzhaf 2007; Crossman et al. 2013; Fisher et al. 2009; Gómez-Baggethun and Barton 2013), among others. The proposed framework refers to the widely applied Common International Classification of Ecosystem Services (CICES V4.3), developed under the auspices of the European Environmental Agency (EEA 2013).

Figure 2.2 schematically represents how urban water infrastructures spatially link the areas of ecosystem services production (i.e. watershed boundary) and benefit (i.e. city boundary). It allows distinguishing between the different roles of upstream and downstream watersheds or comparing the direction of flow of different ecosystem services with that of water. Noteworthy is the spatial mismatch between areas of ecosystem services supply and benefit, and its implications for the sharing of benefits and costs for maintaining ecosystems and their services. For example, important regulating ecosystem services (such as soil erosion control) that determine the quantity and quality of water available for the city are provided by areas in the upstream

Table 2.1 The main ecosystem services intercepted by the urban water sector, classified according to the CICES V4.3

Section[a]	Division[b]	Group[c]	UWI
1. Provisioning	1.1 Nutrition	1.1.1 Water (surface/ground)	WSS
	1.2 Materials	1.2.1 Water (surface/ground water)	WSS
2. Regulation and maintenance	2.1 Mediation of waste, toxics and other nuisances	2.1.1 Mediation by biota remediation/filtration/sequestration	WSS-SS
		2.1.2 Mediation by ecosystem	WSS-SS
	2.2 Mediation of flows	2.2.1 Liquid flows (hydrological cycle/flooding)	WSS-SS-DS

Source EEA (2013)
[a]Categories of ecosystem services, according to the CICES V4.3
[b]Section categories by types of output or process
[c]Division categories by biological, physical or cultural type or process
UWI: Urban water infrastructure; WSS: Water supply system; SS: Sanitation system; DS: Drainage system

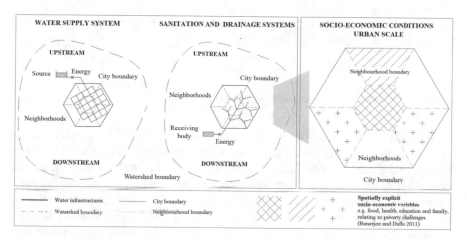

Fig. 2.2 Left: Schematic spatial representation of the role of urban water infrastructures in the flow of ecosystem services from areas of production (watershed boundary) to areas of benefit (city boundary). Right: Zoom on the urban scale showing socio-economic differences between neighbourhoods. (*Source* Adem Esmail 2016)

watershed whereas the main beneficiaries are economic activities and households in the city. Moreover, the scheme in Fig. 2.2 highlights the disparities between different neighbourhoods in terms of their access to benefits from ecosystem services because of an uneven infrastructural coverage (i.e. town system and on-site system) and more generally due to the underlying socio-economic conditions, including issues of poverty and equity (Daw et al. 2011). In this regard, to gain a more nuanced understanding of the lives of the poor, Banerjee and Duflo, for example, suggest application of Randomized Controlled Trials focusing on four key aspects relating to food, health, education and family in order to rethink the way to fight poverty (Banerjee and Duflo 2011).

In general, all the aspects mentioned above lead to crucial decision-making questions that can be framed from a perspective of equity. As part of the fight against poverty, equity is linked in fact to the availability and changes in ecosystem services (MA 2005). A complex and ambiguous concept, equity is here characterized rather simplistically along two fault lines: a geographical (spatial) fault line, related to the distribution of people in the territory (e.g. watershed versus city), and a social fault line, related to how sub-groups within a community are formed (Abebe et al. 2008). Moreover, in the pursue of equity as a societal goal, a pragmatic approach is to ensure that decision-making relies on sufficiently disaggregated analysis (Daw et al. 2011; Ernstson 2013). Disaggregation concerns both ecosystem services, to understand trade-offs, and beneficiaries, to identify losers and winners (Geneletti et al. 2020). While the extent of disaggregation is dependent on the level of existing inequities: the higher the inequities the more disaggregated the analysis (Daw et al. 2011). Therefore, in the proposed framework of the urban water sector, we included

both town system and on-site facilities, recalling the underlying opposing trends and
related equity issues (see Fig. 2.1).

2.1.4 Ecosystem-Based Responses

The last part of the framework in Fig. 2.1 presents an illustrative set of ecosystem-
based responses to mitigate the drivers of water vulnerability, i.e. risks of flooding,
drought and water scarcity. The assumption here is that water vulnerability and
a related concept of natural water variability are representative of the diverse challenges facing the urban water sector; whereas ecosystem-based responses are selected
as cost-effective measures to face water vulnerability (EEA 2012; Vanneuville et al.
2012; Werner and Collins 2012). Water vulnerability is the exposure of water ecosystems and society to shortages and excesses of water caused by humans, taking place
in the form of risks of flooding, droughts and water scarcity (Vanneuville et al. 2012).
Water variability is the variation of the water content that occurs according to the
seasons, geography of the region and the types of water bodies; it takes place in the
form of droughts and flooding (Vanneuville et al. 2012). Therefore, ecosystem-based
responses to drivers of water vulnerability are considered the best way to improve
water quality and minimize water scarcity and floods (Werner and Collins 2012). As
put by Vanneuville and colleagues, they "ensure that healthy ecosystem are able to
function as habitats for a rich biodiversity and, at the same time, are able to retain
water in a natural way and help regulate the hydrological cycle, purifying and filter-
ing water to provide humans and nature with enough clean water" (Vanneuville et al.
2012).

Table 2.2 presents the ecosystem-based responses included in the framework. An
interesting subset is represented by the so-called "natural water retention measures"
(NWRM). They consist of measures that aim to reestablish the natural water variation
by acting on land use and resulting land cover, which are the main factor affecting
the provision of ecosystem services (see Table 2.3 for some examples in different
contexts).

2.2 Relationship with Other Frameworks

As briefly described hereafter, the proposed framework of the urban water sector is connected to other frameworks in the different specific fields covered by its
components.

Ecosystem services frameworks. The proposed conceptual framework considered the two most consolidated ecosystem services frameworks: the *"Millennium
Ecosystem Assessment"*, MA (2005) and the Cascade Model of *"The Economics
of Ecosystems and Biodiversity"*, TEEB (Braat and de Groot 2012; de Groot et al.
2010; Haines-Young and Potschin 2010). The MA serves to identify the ecosystem

Table 2.2   Ecosystem-based response to drivers of water vulnerability

Floods risks	Drought & water scarcity
R1. Restrictions to land use: favour natural water retention measures (NWRMs), by restoring wetlands, increasing forest cover, enhancing natural features of floodplains, reducing impervious surfaces in cities	R1. Restrictions to land use: favour natural water retention measures (NWRMs)
R2. Knowledge and governance: assessing the natural water variability (concept of "flow regime"); adopt a risk management rather than a crisis management approach. The former accepts the occurrence of flooding and drought, but tries to mitigate their effects with preventive action, at a relatively lower societal cost	R2. Knowledge and governance: knowing at any given time and location what water is available for human use and for ecosystem.
	R3. Efficiency: increase water efficiency at the household, industry and irrigation
	R4. Pricing and economic measures: water pricing and metering to change consumption style, taxes and subsidies to discourage water use in certain places and times, thus allocate water resources between competing sectors
	R5. Increase water supply: synergies with the other sectors to reduce pressure (e.g. increase efficiency of irrigation systems, fight against invasive alien plants

Source Burek et al. (2012), Schmidt and Benítez-Sanz (2012), Vanneuville et al. (2012)

Table 2.3   Examples of natural water retention measures in different contexts

Urban	Agricultural	Forest	Water storage
Filter strips and swales	Restoring, maintaining meadows and pastures	Continuous cover forestry	Basins and ponds
Permeable surfaces and filter drains	Buffer strips	Maintaining, developing riparian forests	Wetland restoration and creation
Infiltration devices	Soil conservation crop practices	Afforestation of agricultural land	Floodplain restoration
Green roofs	No or reduced tillage		Re-meandering
	Green cover		Restoration of lakes
	Early sowing		Natural bank stabilization
	Traditional terracing		Artificial groundwater recharge

Source Vanneuville et al. (2012)

services and the constituents of human wellbeing involving the urban water sector. The TEEB is useful to highlight the role of institutions in determining the use of ecosystem services, managing feedbacks between natural and human systems. As shown in Fig. 2.3, the two frameworks can be coupled with schemes of urban water infrastructures, identifying the components that interface with ecosystem services (e.g. the "source" in a water supply system).

For instance, Fig. 2.3 illustrates how the urban water sector (both infrastructures and institutions) plays a crucial role in the flow of ecosystem services to people in cities. Urban water infrastructures, by physically linking areas of ecosystem services production and benefit (i.e. watersheds and cities, respectively), allow the flow of important provisioning and regulating ecosystem services that underpin human wellbeing in terms of health, security and livelihood. Water utilities play a key role in managing the links and feedbacks between cities, infrastructures and watersheds, i.e. between socio-technical and ecological systems. As central actors, they are in a position to affect the feedback between value, benefit and use of water-related ecosystem services (socio-technical side) and to deal with the management and restoration of ecosystem (ecological side).

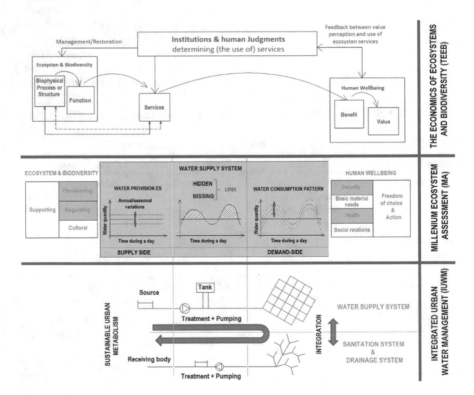

Fig. 2.3 Integrating the urban water sector within MA and TEEB ecosystem services frameworks and IUWM's concept of "sustainable urban metabolism". Example of the water supply system (grey Box). (*Source* Adem Esmail 2016)

More specifically, the grey box in Fig. 2.3 refers to the case of the water supply systems. It illustrates their key role in linking and balancing the water demand and supply sides, at daily, seasonal and annual temporal scales. On a daily basis, water provision at the source can be assumed constant, with consumption patterns reflecting the prevailing socio-economic and cultural habits of the users (e.g. morning and evening peaks); whereas water provision and consumption may have significant annual or seasonal fluctuations. However, this important contribution human well-being is often unacknowledged, especially by end-users, because water infrastructures tend to be "hidden" underground (the case of rich countries from the Global North) or worse "missing" (the case of poor countries from the Global South), and generally "weak" due to ageing and leaking infrastructures and poor institutional arrangements.

Integrated urban water management. The proposed framework also builds on the concept of Integrated Urban Water Management (IUWM). This is an approach that has been proposed to address diverse challenges facing urban water sector (Bahri 2012). Among others, underlying the IUWM is the concept of "sustainable urban metabolism" as opposed to an "unbalanced urban metabolism" (e.g. cities simply importing freshwater from the watershed and releasing wastewater) (Bahri 2012; Cowie and Borrett 2005). Put simply, IUWM advances an integrated management of the whole water cycle within the city. In this regard, in Fig. 2.3, noteworthy are two blue arrows illustrating the concept of sustainable urban metabolism and the pragmatic need for a good understanding and integration of the different water systems.

SEEA-Water framework. Developed by the United Nations Statistics Division (UNSD), the System of Economic and Environmental Accounting of Water (SEEA-Water) consists of standardized concepts and methods in water accounting (UN-DESA 2011). It allows organizing economic and hydrological information, enabling a consistent analysis of the contribution of water to the economy and of the impact of the economy on water resources. Therefore, the proposed framework in Fig. 2.1 is structured in accordance with the conceptual foundation of the SEEA-Water. For example, a distinction is made between upstream and downstream watersheds, which is crucial for properly framing the diverse challenges they face. On the other, for ease of application and flexibility, in the framework, the use of layperson terminology (e.g. households, commercial users, roads and sidewalks) is preferred to the standardized sector codes of the SEEA-Water, based on the "International Standard Classification of all Economic Activities" (ISIC Rev.4). Thus, the proposed framework is both intuitive (e.g. represents the urban water cycle using layperson terms and arrows) and flexible (e.g. can be easily adapted to meet the context-specific needs and the desired levels of detail and complexity). At the same time, relying on the conceptual foundation of the SEEA-Water, it easily draws from its rich set of indicators and methods.

2.3   An Example of Application

The proposed framework was used to review real-life projects of urban water infrastructures. Based on the selection criteria shown in Fig. 2.4, five major projects, funded by the World Bank, during the years 2003-2013 were considered. The projects are all located in Eastern Africa, an area characterized by several water-related socio-ecological challenges and a high demand for new infrastructures. For each project, three types of documents were reviewed: (i) *project paper*, (ii) *EIA document* and (iii) *resettlement plans* (Table 2.4). The proposed conceptual framework was used to reorganize systematically the information from the project documents; hence, a binary scoring system was applied to assess the coverage of each part of the framework.

The five reviewed World Bank projects are listed in Table 2.5. All the projects had both an "infrastructural" and an "institutional and capacity building" component, which was here overlooked. They all addressed a single urban water infrastructure, except for the case of Addis Ababa and Blantyre, which jointly consider water supply and sanitation and drainage system. The five projects mainly dealt with town systems; yet, on-site systems were mentioned in the case of Addis Ababa (e.g. "in high-income residential areas sanitation will be based on on-site septic tank systems financed by the owner") and Blantyre (e.g. "construction of 100 kiosks"). The five projects provide detailed information about the project beneficiaries (indirectly, ecosystem services beneficiaries), specifying their present and future demands. The Maputo project is the only one that mentioned import of water from other watersheds. In general, the concept of ecosystem services was not expressly used as a holistic framework; yet, water-provisioning ecosystem services were more clearly identified than the regulating services. Only the Kampala project explicitly mentioned the use of ecosystem-based enhancement of tertiary treatment of effluent, by restoring wetlands. While the Malawi project included the establishment of a pilot Catchment Management Authority to promote, among others, the "preservation and enhancement of key environmental systems".

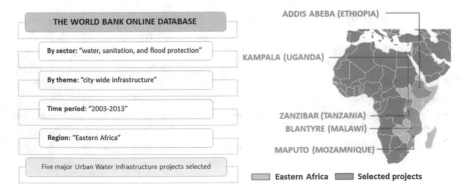

Fig. 2.4 Search and selection of real-life water infrastructure projects for review. Sector is "a high-level grouping of economic activities based on types of goods or services produced"; theme is "the pursued goal and priority". (Modified after Adem Esmail 2016)

Table 2.4 Types of reviewed project documents from the World Bank Database

Reviewed document	
1. Project paper and information	*What to look for?*
Country and Sector background	• Significant water-related challenges (flooding and water scarcity) • Other drivers of water vulnerability in cities (population growth, poverty, natural resources conservation and climate change)
Project development objectives	• Strategies to face water challenges in order to meet general development needs • Strategies to achieve long-term human wellbeing without depleting the sustaining ecosystem services
Project description	• Ecosystem services-based actions included in the projects
2. EIA document	*What to look for?*
	• Degree of awareness of the impact to society and ecosystem • Insight on trade-offs and synergies identified as most significant, and how they are dealt with
3. Resettlement plans	*What to look for?*
	• Insight on how ecosystem services production and benefit area mismatch is addressed

However, an in-depth analysis of the urban water sector from an ecosystem services perspective, as originally intended, was not possible based on the information included in the reviewed documents. Additional sources would have been needed to reach more nuanced conclusions about the urban water sector in the five selected cities. Still, the review provides insights about the key aspects synthesized in the proposed framework considered in the water infrastructure projects.

2.4 Concluding Remarks

The aim of this chapter was broad and challenging, involving multiple systems and concepts, for example, water supply and sanitation systems, water governance, engineering design, water accounting and ecosystem services. It attempted to represent the prevailing management paradigms, i.e. "a set of basic assumptions about the nature of the system to be managed, the goals of managing the system and the ways in which these goals can be achieved, shared by an epistemic community of actors involved in the generation and use of relevant knowledge" (Pahl-Wostl et al. 2011). Therefore, the findings and the proposed framework are arguable, not exhaustive and do not reach the level of detail that would be needed to gain a nuanced understanding of the urban water sector. Nevertheless, despite these limitations, they can be a useful starting point for seeking a better understanding of the complex relationship

Table 2.5 Five water infrastructure projects from the World Bank Online Database, reviewed using the proposed framework (Project location, budget, title and development objective)

Location	Project title	Project Development objective	Project budget
Addisa Ababa (Ethiopia)	Urban water supply and sanitation	(a) to produce and distribute more water and improve sanitation services in Addis Ababa and other targeted secondary cities, (b) to improve operational efficiency (…) and (c) to improve governance by the water boards and to introduce performance incentives for operators	$100 +$85 million
Kampala (Uganda)	Kampala institutional and infrastructure development adaptable program	Develop a strong governance and institutional structure () to enhance service delivery and improve the economic performance of Kampala, through: (a) a program of institutional and fiscal reform (..) and (b) a program of investment at the city-wide scale, focusing on the areas of drainage, roads/traffic management and solid waste removal	$33.6 million
Blantyre (Malawi)	Second national water development project	(a) increase access to sustainable water supply and sanitation services for persons living in cities, towns, villages and Market Centers within the Recipient's territory and (b) improve water resources management at the national level	$120 million

(continued)

Table 2.5 (continued)

Location	Project title	Project Development objective	Project budget
Zanzibar (Tanzania)	Zanzibar urban services project	Improve access to urban services in Zanzibar and conserve the physical cultural heritage at one public location within the "Stone Town"	38 million
Maputo (Mozambique)	Greater Maputo water supply expansion	Increase access to clean water for residents in the Greater Maputo Area	$178 million

between long-term human wellbeing in cities and the respective service providing and life-supporting watersheds.

The proposed framework synthesizes the most relevant aspects characterizing the exchange of water between watersheds and cities, and within the city. It highlights the role of urban water infrastructures in (i) linking ecosystem services production and benefit areas, (ii) bridging spatial scales ranging from the watershed to the household level and (iii) adopting ecosystem-based responses to water vulnerability. A possible application of the framework is to use it as a tool for reviewing infrastructural projects. By reorganizing the information, it is possible to assess the extent to which the different aspects of the framework (e.g. ecosystem-based responses, IUWM) have actually been taken into account. The framework sets a useful background for further analysis of the urban water sector, as presented in the following chapter focusing on the watershed scale.

References

Abebe H, Bedru M, Ashine A, Hilemeriam G, Haile B, Demtse D, Adank M (2008) Equitable water service for multiple uses: a case from Southern Nations nationalities and peoples region (SNNPR), Ethiopia (No. 17). Dire Dawa, Ethiopia

Adem Esmail B (2016) Ecosystem services for watershed management and planning (PhD thesis). University of Trento, Italy

Bahri A (2012) Integrated urban water management. TEC Background Papers

Banerjee AV, Duflo E (2011) Poor economics. In: A radical rethinking of the way to fight global poverty. PublicAffairs, New York. https://doi.org/10.3362/1755-1986.2011.037

Boyd J, Banzhaf S (2007) What are ecosystem services? The need for standardized environmental accounting units. Ecol Econ

Braat L, de Groot RS (2012) The ecosystem services agenda: bridging the worlds of natural science and economics, conservation and development, and public and private policy. Ecosyst Serv 1:4–15. https://doi.org/10.1016/j.ecoser.2012.07.011

Burek P, Mubareka S, Rojas R, Roo D, Bianchi A, Baranzelli C, Lavalle C, Vandecasteele I (2012) Evaluation of the effectiveness of natural water retention measures support to the EU blueprint to safeguard Europe' s. Luxembourg. https://doi.org/10.2788/5528

Choguill CL (1996) Ten steps to sustainable infrastructure. Habitat Int. 20:389–404. https://doi.org/10.1016/0197-3975(96)00013-6

Choguill CL (1999) Community infrastructure for low-income cities: the potential for progressive improvement. Habitat Int. 23:289–301. https://doi.org/10.1016/S0197-3975(98)00053-8

Cowie GM, Borrett SR (2005) Institutional perspectives on participation and information in water management. Environ Model Softw 20:469–483. https://doi.org/10.1016/j.envsoft.2004.02.006

Crossman ND, Burkhard B, Nedkov S, Willemen L, Petz K, Palomo I, Drakou EG, Martín-Lopez B, McPhearson T, Boyanova K, Alkemade R, Egoh B, Dunbar MB, Maes J (2013) A blueprint for mapping and modelling ecosystem services. Ecosyst Serv 4:4–14. https://doi.org/10.1016/j.ecoser.2013.02.001

Daw TM, Brown K, Rosendo S, Pomeroy R (2011) Applying the ecosystem services concept to poverty alleviation: the need to disaggregate human well-being. Environ Conserv 38:370–379. https://doi.org/10.1017/S0376892911000506

de Groot RS, Alkemade R, Braat L, Hein L, Willemen L (2010) Challenges in integrating the concept of ecosystem services and values in landscape planning, management and decision making. Ecol Complex 7:260–272. https://doi.org/10.1016/j.ecocom.2009.10.006

EEA (2012) European waters—assessment of status and pressures. European Environment Agency. https://doi.org/10.2800/63266

EEA (2013) The common international classification of ecosystem services—Version 4.3 (CICES)

Ernstson H (2013) The social production of ecosystem services: a framework for studying environmental justice and ecological complexity in urbanized landscapes. Landsc Urban Plan 109:7–17. https://doi.org/10.1016/j.landurbplan.2012.10.005

Fisher B, Turner RK, Morling P (2009) Defining and classifying ecosystem services for decision making. Ecol Econ 68:643–653. https://doi.org/10.1016/j.ecolecon.2008.09.014

Geneletti D, Cortinovis C, Zardo L, Adem Esmail B (2020) Towards equity in the distribution of ecosystem services in cities. In: Geneletti D, Cortinovis C, Zardo L, Adem Esmail B (eds) Planning for ecosystem services in cities, SpringerBriefs in environmental science. Springer International Publishing, Cham, pp. 57–66. https://doi.org/10.1007/978-3-030-20024-4_6

Gómez-Baggethun E, Barton DN (2013) Classifying and valuing ecosystem services for urban planning. Ecol Econ 86:235–245. https://doi.org/10.1016/j.ecolecon.2012.08.019

Haines-Young R, Potschin M (2010) The links between biodiversity, ecosystem services and human well-being, In: Raffaelli D, Frid C (eds), D. Raffaelli/C. Frid (Hg.) Ecosystem ecology: a new synthesis. Cambridge University Press, pp 110–139. https://doi.org/10.1017/CBO9780511750458

Larsen TA, Gujer W (1997) The concept of sustainable urban water management. Water Sci Technol 35. https://doi.org/10.1016/S0273-1223(97)00179-0

Lieberherr E, Truffer B (2015) The impact of privatization on sustainability transitions: a comparative analysis of dynamic capabilities in three water utilities. Environ Innov Soc Trans 15:101–122. https://doi.org/10.1016/j.eist.2013.12.002

MA (2005) Ecosystems and human well-being: synthesis. In: A report of the millenium ecosystem assessement. Island Press, Washington, DC

Pahl-Wostl C, Jeffrey P, Isendahl N, Brugnach M (2011) Maturing the new water management paradigm: progressing from aspiration to practice. Water Resour Manag 25:837–856. https://doi.org/10.1007/s11269-010-9729-2

Richter BD, Abell D, Bacha E, Brauman K, Calos S, Cohn A, Disla C, O'Brien SF, Hodges D, Kaiser S, Loughran M, Mestre C, Reardon M, Siegfried E (2013) Tapped out: how can cities secure their water future? Water Policy 15:335–363. https://doi.org/10.2166/wp.2013.105

Schmidt G, Benítez-Sanz C (2012) Topic report on: assessment of water scarcity and drought aspects in a selection of European Union River Basin Management Plans

Suleiman L et al (2019) Diverse pathways—common phenomena: comparing transitions of urban rain harvesting systems in stockholm, Berlin, Barcelona. J Environ Plan, Manag

Tilley E, Lüthi C, Morel A, Zurbrügg C, Schertenleib R (2014) Compendium of sanitation systems and technologies, 1st edn, Development. Swiss Federal Institute of Aquatic Science and Technology (Eawag), Duebendorf, Switzerland. https://doi.org/SAN-12

UN-DESA (2011) System of environmental-economic accounting for water (SEEA-Water)

Vanneuville W, Werner B, Kjeldsen T, Miller J, Kossida M, Tekidou A, Kakava A, Crouzet P (2012) Water resources in Europe in the context of vulnerability: EEA 2012 state of water assessment. https://doi.org/10.2800/65298

Werner B, Collins R (2012) Towards efficient use of water resources in Europe. https://doi.org/10.2800/95096

WHO, UN-HABITAT (2010) Hidden cities: unmasking and overcoming health inequities in urban settings

World Bank (1994) World development report: infrastructure for development. Washington, DC

Chapter 3
An Operational Approach for Watershed Investments

Abstract This chapter focuses on Watershed Investments for securing water for cities. It starts with a brief account of the application of ecosystem services for decision-making and a theoretical background of boundary work. Accordingly, it proposes an operational approach developed for designing and assessing impact of watershed iInvestments to secure water for cities. The developed approach distinguishes between a "strategic" and a "technical" component. The strategic component identifies as key inputs of the process of Watershed Investment design and assessment, the definition of objectives and visioning of feasible and desirable scenarios by stakeholders. The technical component applies spatially explicit modelling to design Watershed Investments, hence to model the impacts on selected ecosystem services. The chapter concludes highlighting the potential of the approach to contribute to adaptive management in the urban water sector, by addressing the challenges of linking diverse stakeholders and knowledge system across management levels and institutional boundaries.

Keywords Ecosystem services · Integrated urban water management · Adaptive urban water management · Boundary work · Water Funds · Water security · Sub-Saharan Africa

3.1 Watershed Investments for Ecosystem Services: A Promising Opportunity

Watershed Investments—also referred to as Water Funds or Investment in Watersheds Services—consist of governance and financial mechanisms that secure clean water for downstream users, mainly cities, and operate by engaging primarily upstream communities and nature conservation organizations (Adem Esmail and Geneletti 2017; Hamel et al. 2020; Higgins and Zimmerling 2013). They target a wide range of activities, from changes in land use and alteration of vegetative covers to education and community outreach, enhancing specific ecosystem services such as erosion control and nutrient retention, while conserving nature and biodiversity. Watershed Investments can also have explicit social objectives such as poverty alleviation, here understood as both poverty reduction and prevention (Daw et al. 2011).

   Watershed Investments are acknowledged as a promising opportunity to promote large-scale transformative change that promotes human wellbeing while conserving ecosystems (Guerry et al. 2015; Kroeger et al. 2019; McDonald and Shemie 2014). Based on an in-depth analysis of watersheds supplying 500 cities worldwide, 25% of the analysed cities would gain a positive return from Watershed Investments, with annual saving on water treatment costs exceeding US$ 890 million (McDonald and Shemie 2014). Designing and assessing Watershed Investments, however, can be challenging. Among others, it has to deal with a diverse set of barriers to transfer of knowledge into action and the related boundary work concerns. In this regard, there is a need for adequate approaches for supporting the design and implementation of Watershed Investments, by duly addressing the concerns of different stakeholders. This includes taking into account both contextual and contingent factors as well as the relative influence of stakeholders, broadly defined here as any actor that is affected by and/or can affect Watershed Investments, such as upstream communities, downstream users, water management agencies, conservationists and funding bodies.

   From a decision-making perspective, a key aspect of Watershed Investments deals with the fact that—through change of land cover—they can also serve to enhance the supply of ecosystem services at a scale of basins. Indeed, consideration of the effects on ecosystem services is increasingly included in decision-making (Abson et al. 2014; Burkhard and Maes 2017; de Groot et al. 2010; Haase et al. 2014) and impact assessment processes (Geneletti 2015; Geneletti et al. 2016; Mandle et al. 2016). Spatially explicit modelling of ecosystem services, in particular, allows generating and exploring future scenarios, to understand the trade-offs between different Watershed Investments' objectives and to possibly optimize co-benefits—for instance, by exploiting urban nexuses (GIZ and ICLEI 2014) and synergies between ecosystem services (Howe et al. 2014). Good examples that apply spatially explicit ecosystem services modelling for tradeoff analysis can be found in (Geneletti 2013; Geneletti et al. 2018; Lawler et al. 2014; Polasky et al. 2008), to name a few. These studies demonstrate how a set of designed land use patterns can better meet competing objectives, such as nature conservation, agricultural production and urban growth.

   This chapter proposes an operational approach for designing and assessing the impacts of Watershed Investments, considering their unintended, and typically unattended, consequences both within and beyond the watershed. The proposed approach combines spatially explicit modelling of ecosystem services with insights on boundary work as a strategy to overcome barriers of transfer of knowledge into action. Sections 3.2 and 3.3 introduce the application of ecosystem services for decision-making and an account the theoretical background of boundary work, respectively. These two concepts in fact lay the foundations for proposed operation approach. The remainder of the chapter presents a detailed description of the proposed operational approach for designing and assessing impacts of Watershed Investments, specifying the rationale behind each step of the process.

3.2 Boundary Work: Theoretical Background

The concept of boundary work, a subset of the bridging organization literature (Berkes 2009; Crona and Parker 2012; Olsson et al. 2007), was originally introduced to understand efforts to demarcate science from non-science (Gieryn 1983). Since then, the concept has evolved to provide a better framing of an active management of the tension that arises at the interface between stakeholders involved in knowledge co-production. An example of application of the boundary work concept for implementing adaptive management in the urban water sector is reported in Adem Esmail et al. (2017). For the scope of this book, in the following, we present the main elements of the boundary work framework as synthesized framework in Fig. 3.1.

The most innovative aspect of the framework by Clarck and colleagues is its systematic classification into nine different contexts of knowledge co-production, based on what knowledge is used for and how users perceive its source. This makes the framework highly effective in capturing potential barriers, hence in identifying the most appropriate boundary work strategies (i.e. bundle of specific actions and measures) that could be deployed to overcome them. The classification considers three types of knowledge uses and three types of knowledge sources. In terms of use, knowledge can contribute to enlightenment, support decision-making by a single user or support negotiations between multiple users (columns in Fig. 3.1). In terms of

BOUNDARY WORK		USE of knowledge to support...		
		A. Enlightenment (U0)	B. Decision (U1)	C. Negotiation (Um)
		CREDIBILITY	CREDIBILITY + SALIENCY	CREDIBILITY+SALIENCY+LEGITIMACY
SOURCE of knowledge...	Personal expertise (S0)	$S_o \leftrightarrow U_o$ *Contemplation*	$S_i \leftrightarrow U_j$ *Decision*	$S_i \leftrightarrow \updownarrow \begin{matrix} U_k \\ U_\ell \end{matrix}$ *Politics*
	Single community of expertise (S1)	$S_1 \leftrightarrow U_o$ *Demarcation* A.2	$S_i \leftrightarrow U_j$ *Expert advice* B.1 B.2	$S_i \leftrightarrow \updownarrow \begin{matrix} U_k \\ U_\ell \end{matrix}$ *Assessment* C.1
	Multiple communities of expertise (Sn)	$\begin{matrix} S_i \\ \updownarrow \leftrightarrow U_o \\ S_j \end{matrix}$ *Integrative R&D* A.1	$\begin{matrix} S_i \\ \updownarrow \leftrightarrow U_j \\ S_j \end{matrix}$ *Participatory R&D*	$\begin{matrix} S_i \quad U_k \\ \updownarrow \leftrightarrow \updownarrow \\ S_j \quad U_\ell \end{matrix}$ *Political bargaining*

Fig. 3.1 A generalized framework of boundary work (modified after Clark et al. 2016)

source, users may perceive knowledge as originating from themselves, from a single community of expertise or from multiple and potentially conflicting communities of expertise (rows in Fig. 3.1).

For each context, based on the analysis of decades of international agricultural research, Clark et al. (2016) provide examples of barriers to knowledge co-production as well as define the most appropriate boundary work strategy to overcome them. Altogether, they identify nine strategies, which they label as: *Contemplation, Decision, Politics, Demarcation, Integrative Research & Development, Expert Advice, Participatory Research & Development,* and *Assessment and Political Bargaining.* These strategies span from the simplest context in which knowledge is used for enlightenment and stakeholders perceive it as their own (*Contemplation*), to the most challenging context in which stakeholders with divergent interests use knowledge from potentially conflicting sources, for negotiation purposes (*Political bargaining*). Here, it suffices to note these strategies differ in terms of the effectiveness criteria that have to take into account.

In fact, for each context and related strategy, the framework by Clark and colleagues defines the effectiveness criteria of boundary work to be considered. Three criteria determine the effectiveness of boundary work: *credibility, saliency and legitimacy. Credibility* refers to technical and scientific adequacy in the handling of evidence, *saliency* to the relevance to the problem at hand and *legitimacy* to the fact that boundary work ought to be fair, unbiased and respectful of all stakeholders (Cash et al. 2003). The effectiveness criteria to consider depend mainly on the type of knowledge use and less on the source. Thus, for *enlightenment*, boundary work should only ensure *credibility*; for *decision-making*, it should achieve both *credibility* and *saliency*; finally, for *negotiation*, it should jointly consider trade-offs between *credibility, saliency* and *legitimacy* (columns' title in Fig. 3.1).

According to Clark et al. (2016), three attributes generally increase the likelihood of success of boundary work, regardless of the context. Namely, participation of stakeholders in agenda setting and knowledge production (i.e. participation), governance arrangements that assure accountability of the resulting boundary work to relevant stakeholders (i.e. accountability) and collaborative products, such as maps and reports, that are adaptable to different viewpoints (i.e. boundary objects) (Star and Griesemer 1989). Concerning the functions that most contribute to successful knowledge co-production, Cash et al. (2003) identify three, namely, iterative and inclusive communication (i.e. communication), translation of concepts to facilitate mutual understanding (i.e. translation) and mediation efforts to resolve potential conflicts (i.e. mediation).

Noteworthy, boundary work is not a single-time achievement rather is a dynamic process that has to address an entire "landscape of tensions" (Parker and Crona 2012). Tension refers to the interface between stakeholders engaged in knowledge co-production, including tension between "basic vs applied research", "disciplinary vs interdisciplinary", "long-term vs real-time" and "autonomy versus consultancy" (Parker and Crona 2012). Thus, boundary work has to be conceived as dynamic process taking place within a given embedding socio-ecological system. A good understanding of the contextual (relatively stable, e.g. regional identity) and contingent (relatively changeable, e.g. technological innovation) factors and the relative influence of social actors and their role in knowledge co-production remains thus a prerequisite for any boundary work.

3.3 An Approach for Designing and Assessing Impacts of Watershed Investments

Based on the ecosystem services and boundary work considerations, we propose an operational approach for designing and assessing impact of Watershed Investments. By taking into account the different stakeholders concerns and related boundary work needs, the proposed approach distinguishes between a strategic component (mainly addressing *saliency* and *legitimacy*) and a technical component (mainly addressing *credibility*) (Fig. 3.2). Each component is divided into an initial, intermediate and final stage, reflecting the dynamic nature of boundary work and the important role of timing. The three stages of the strategic component are described in Sect. 3.4 and include setting the agenda, defining investment scenarios and assessing their performance to then plan for a follow-up. The technical component, described in Sect. 3.5, includes three main stages: biophysical data processing, tailoring of spatially explicit ecosystem services models, followed by their application to generate

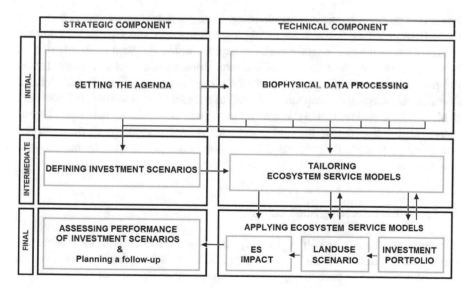

Fig. 3.2 Proposed operational approach for designing and assessing impact of Watershed Investments: two-component and three-stage, process-based approach, building on the concepts of ecosystem services and boundary works (*Source* Adem Esmail and Geneletti 2017)

investment portfolios and land use scenarios and to model the impacts on a set of selected ecosystem service.

3.4   The Strategic Component

3.4.1   Setting the Agenda

This initial stage requires a deep understanding of the socio-ecological context in which the proposed approach is to be applied, including the contextual and contingent factors and the relative influence of the social actors. From a boundary work perspective, joint agenda setting can help achieve *saliency* and *legitimacy*, and thus contribute to building an "an arena for knowledge co-production, trust building, sense-making, learning, vertical and horizontal collaboration and conflict resolution" (Berkes 2009). An agenda jointly set by the stakeholders can be a first important boundary object of the process of Watershed Investment design and assessment. In our approach, setting the agenda consists of the following four steps.

3.4.1.1   Defining Objectives and Planning Horizon, Identifying and Prioritizing Ecosystem Services

Numerous factors contribute to defining Watershed Investment objectives and the related planning horizon, including legal limitations and standards, empirical evidence, stakeholders' negotiations and experts' opinions (Vogl et al. 2015). This is followed by the identification of the most relevant ecosystem services, hence, the prioritization and selection of some for more in-depth analyses (i.e. Watershed Investments design and impact assessment), as suggested, for example, in Landsberg et al. (2013), Geneletti (2015). What is important is that defining Watershed Investments' objectives, identifying and prioritizing related ecosystem services provide a crucial opportunity for enhancing both *saliency* and *legitimacy*, by involving as much as needed the different stakeholders.

3.4.1.2   Characterizing Stakeholders

A promising concept useful to identify and characterize stakeholders is that of service-sheds (Tallis et al. 2015). It consists in ideally tracking the flow of ecosystem services that benefit a particular group of people, hence, delimitating the geographic areas of production of such ecosystem services. Service-sheds can be an effective tool for highlighting key societal issues such as the spatial mismatch between areas of ecosystem services production and benefit, sharing of costs for maintaining ecosystem services and more generally addressing challenges of inter and intra-generational equity (Lamorgese and Geneletti 2015). However, tracking flow of ecosystem services can be quite cumbersome (Pagella and Sinclair 2014); especially when considering the different ways in which groups of people with different socio-economic conditions benefit from ecosystem services (Daw et al. 2011). An additional challenge for Watershed Investments is to distinguish between "losers" and "winners". The former comprise those negatively affected by activities undertaken in the frame of the Watershed Investments (e.g. farmers losing access to protected areas or introducing new agricultural practices). The latter include those whose wellbeing the Watershed Investments considers as a target (e.g. people in cities benefiting from more water or rural communities involved in poverty alleviation initiatives). However, the two categories are not mutually exclusive, such as the case of rural communities that lose access to protected area, but at the same time, benefit from poverty alleviations measures. Therefore, defining, in a more detailed way, "who is who" and their role in the stakeholders' negotiation is left as an empirical question during the actual application of the proposed approach. In turn, this determines the type of boundary work that needs to be put in place, in order to address the concerns of *saliency* and *legitimacy* of the different stakeholders.

3.4.1.3 Budgeting Watershed Investments

We propose a relatively simple approach, which consists in estimating the value of ecosystem services, by adopting the perspective of a given stakeholder (e.g. loss of storage capacity due to soil erosion, which negatively affects a water utility). To an extent, this choice accounts for a general tendency of stakeholders to strongly prefer avoiding losses to acquiring gains, the so-called "loss aversion" bias (Kahneman and Tversky 1984). A good working example is the case of New York City and the Catskills-Delaware watersheds in which the issue of budgeting was affected strongly and favourably by the adoption of an ecosystem services perspective (Turner and Daily 2008).

3.4.1.4 Selecting Activities

This step requires good contextual knowledge and substantial input from the different stakeholders; all are very important for identifying the activities, specifying their effectiveness and determining their unitary costs. The selected activities need to be characterized adequately in order to determine their potential contribution in meeting the objectives of the Watershed Investment, and this should take into account the preference of the different stakeholders. In fact, activities may diversely involve and affect different stakeholders, for example, measures taken to protect natural areas could limit access to local communities, on the other hand, improvement of agricultural vegetation management could increase crop yield. Therefore, the process of selection and characterization of activities can be an important way of fostering a "meaningful communication" among stakeholders, thus contributing to ensure both *saliency* and *legitimacy*.

3.4.2 Defining Investment Scenarios

For a given agenda, defining investment scenarios aims at addressing key questions, such as how does the budget level affect the outcomes of Watershed Investments? What is the most cost-effective combination of activities, for a given budget? and what if the whole budget is allocated to a single activity? In the proposed approach, scenarios are defined based on three elements: Watershed Investment objective, budget level and budget allocation modality. Two budget allocation modalities are considered: one based on cost-effectiveness of activities and one in which the entire budget is pre-allocated to a single activity at a time. Above all, investment scenarios need to be discriminated based of their desirability and/or feasibility, through a meaningful participation of stakeholders. Jointly defining and discriminating investment scenarios allows further addressing the concerns of *saliency* and *legitimacy* of different stakeholders, also benefiting from their previous interactions in the initial stages of both the strategic and technical components.

3.4.3 Assessing the Performance of Investment Scenarios and Follow-up Planning

The last stage of the strategic component deals with comparing the performance of different investment scenarios based on the findings of the technical component (described in Sect. 3.5) and taking into account the feasibility and desirability of scenarios expressed by the different stakeholders. Ultimately, it aims at selecting, in a participatory fashion, the Watershed Investments to be implemented on the ground; hence, at planning for a follow-up that covers the entire planning horizon. The performance of each investment scenario can be assessed in terms of the change occurring in selected ecosystem services, expressed in their biophysical, social or economic values, with respect to the baseline conditions. Importantly, these aggregated performance values are also supported by spatially and temporally explicit data from the technical component. Such data, besides providing detailed guidance on how to implement the selected Watershed Investment (which activity, where and when), also forms the basis for planning a follow-up (i.e. monitoring, evaluation, management and communication), which is a crucial yet often overlooked phase in the impact assessment practice (Morrison-Saunders et al. 2007). Boundary work wise, the last stage of the strategic component has to ensure *saliency* and *legitimacy* towards the different stakeholders, building on the scientific *credibility* of the findings of the technical component. This is a challenging task because, at times, stakeholders' interests and needs can be simply incommensurable (Parker and Crona 2012), and its success is determined by the extent to which *credibility*, *saliency* and *legitimacy* concerns have been effectively and timely addressed in all the previous stages.

3.5 The Technical Component

3.5.1 Biophysical Data Processing

This stage analyses the need and the processing of biophysical data related to the Watershed Investment objectives. It is thus entirely dependent on the identification and prioritization of key ecosystem services in the strategic component (Sect. 3.4). These are the ecosystem services considered for spatially explicit modelling, to support the Watershed Investments design and assessment. The type of biophysical data and its processing is highly linked to the specific ecosystem services, as will be shown in the case study application in Chap. 5. This stage plays a significant role in ensuring scientific *credibility* of the proposed approach and its outputs, by providing opportunities for linking expert and stakeholders' knowledge. For example, data and resource scarcity, which are perhaps the main challenges, at this stage, could trigger meaningful communication between stakeholders and experts, concerning data collection campaigns, selection of modelling tools and complexity and uncertainty of models, among others.

3.5.2 Tailoring Ecosystem Services Models

Two software tools are at the core of the proposed approach for designing Watershed Investment: RIOS (Vogl et al. 2015) and InVEST (Sharp et al. 2015). RIOS stands for "Resource Investment Optimization System"; it allows targeting investments, based on stakeholders' ecosystem services objectives, their preferences about where activities may occur, and the amount of money that is available for implementing the activities (Vogl et al. 2015). It applies a relative-ranking approach, considering important biophysical factors that drive the specific ecosystem service (Vogl et al. 2015). RIOS is used for designing investment portfolios, where the most cost-effective locations for activities are provided, and generating future land use scenarios. Whereas InVEST is a suite of spatially explicit ecosystem services modelling tools that quantify service provision (Sharp et al. 2015). In the proposed approach, InVEST is used to model the impacts on selected ecosystem services, based on the results of RIOS. Both the tools have been developed by the *Natural Capital Project* during an over-a-decade-long transdisciplinary research, carried out involving real-life stakeholders of Watershed Investments. Thus, from a boundary work perspective, the tools, approaches and lessons learned by this transdisciplinary project (e.g. Ruckelshaus et al. 2015) also

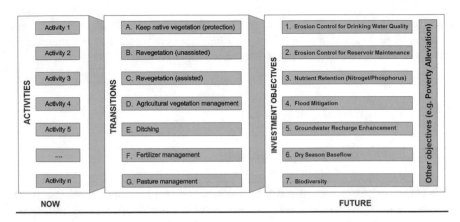

Fig. 3.3 Rationale of the RIOS approach—Investing in watershed activities, now, to trigger transitions in the land use and management, thus meeting investment objectives in the future (*Source* Adem Esmail and Geneletti 2017)

underpin the scientific *credibility* of our approach for designing, and assessing the impacts of Watershed Investment.

Specifically, Figure 3.3 summarizes the rationale behind the RIOS approach described in Vogl et al. (2015). It shows how Watershed Investments target directly a range of activities (now), to trigger a relatively finite set of changes in the watershed, ultimately causing a desired transition in the land use and management (in the future). Such transitions affect many of the processes that regulate hydrologic processes and biodiversity, such as water infiltration rates, soil storage capacity, vegetation cover and structure as well as the maintenance of habitat quality and feeding and breeding resources for species (Vogl et al. 2015). Ultimately, they affect the future land use and related ecosystem services, thus contributing to meeting specific objectives of the Watershed Investment.

Technically, RIOS supports any type of activity at landscape-level, but it does not include grey infrastructure solutions such as check dams and retaining walls. Moreover, each activity must map to one of the seven supported transitions, shown in Fig. 3.3, namely, keeping native vegetation (retaining vegetation likely be lost); assisted or unassisted revegetation (revitalizing vegetation on degraded lands with or without out active interventions); agricultural vegetation management (in creasing crop structure, coverage and/or diversity); ditching (improving infiltration and slowing sediment and nutrients transport); fertilizer management (changing fertilizer application) and pasture management (changing management practices). Likewise, RIOS supports different investment objectives, including seven objectives that are related to ecosystem services and other objectives defined by the user outside of RIOS, such as poverty alleviation (see Fig. 3.3). In the RIOS approach, critical inputs include the so-called transition potentials (i.e. which activities cause which transitions), the objective-transition weights (i.e. relative contribution of each transition to the objective of the Watershed Investment) and the unit cost of each watershed activity. Other important inputs are additional restrictions on watershed activities related, for example, to land use and land cover, slope or elevation. For a full description of the ecosystem services models, including information on their assumptions and limitations, and data pre-processing refer to Sharp et al. (2015) and Vogl et al. (2015).

3.5.3 Applying Ecosystem Services Models

3.5.3.1 Designing Investment Portfolio

For each scenario, the RIOS module "*Investment Portfolio Advisor*" is applied to combine the biophysical and socio-economic input data from the sections above in order to design a set of investment portfolios. An *investment portfolio* consists of a spatially and temporally explicit allocation of the overall budget between the watershed activities. For a given scenario (i.e. Watershed Investment objective, budget level, budget allocation modality), it defines which activities, when during the planning horizon and where in the watershed are the most cost-effective. A simplified description for "non-experts" of the type of data needed and how it is analysed to design the investment portfolio are given in Table 3.1 and Fig. 3.4. From a boundary work perspective, this represents a good example of "*translation*" of concepts to facilitate mutual understanding, aiming at ensuring *credibility* in the eyes of the stakeholders.

Table 3.1 Simplified description of input data type for the RIOS Module "Investment Portfolio Advisor"

Description of input data type	
1	Land use/land cover map
2	Table defining activities and indicating on which land cover types they are allowed
3	Landscape factors influencing effectiveness of transitions to achieve each objective
4	Location and number of beneficiaries benefiting from activities in different areas
5	Factor weights that describe the relative importance of each factor (and process)
6	Objective weights assigning a relative weight when considering multiple objectives
7	Activity-transition table indicating which activities cause which transitions
8	Activity preference areas
9	Floating budget and/or budgets by activity
10	Activity costs

Source Vogl et al. (2015)

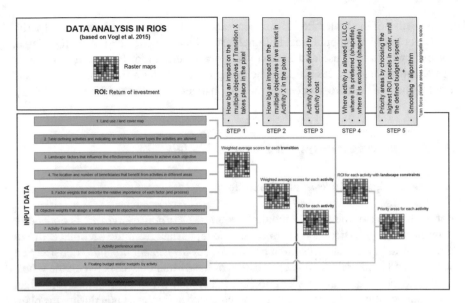

Fig. 3.4 Workflow diagram of the RIOS "Investment Portfolio Advisor": processing of input data, to design Watershed Investments

3.5.3.2   Generating Land Use Scenarios

For each investment portfolio, the RIOS module "*Portfolio Translator*" is applied to generate land use scenarios that are related to the watershed activities. They represent the future condition of the watershed, where RIOS-selected activities are implemented and embedded into the map of Land Use and Land Cover (LULC). Technically, for each existing LULC type, the new LULC that would result from a specific transition needs to be specified. Noteworthy, some transitions (e.g. *assisted revegetation*) imply an actual land use change, while others (e.g. *ditching*) result in a change of the biophysical parameters that affect the ecosystem service of interest. For example, in the case of soil erosion control, such biophysical parameter include sediment retention, sediment export and Universal Soil Loss Equation (USLE) crop factor. An additional RIOS parameter, used to generate future land use, is the "*Proportional Transition Factor—PTF*", which specifies what proportion of the baseline LULC is likely to be transitioned to the new LULC at the end of the planning horizon. Indeed, defining the PTF, which is an important parameter that is linked tightly to the specific socio-ecological system, provides a good opportunity for incorporating local knowledge of stakeholders, and thus underpin the *credibility* of the process.

3.5.3.3 Modelling Impacts on Selected Ecosystem Services

The last stage of the technical component deals with modelling the impacts on a set of selected ecosystem services, using InVEST. The RIOS-generated future land use scenarios and biophysical parameters represent the input data. In the case study application in Chap. 5, for example, the impacts on soil erosion control are modelled, using sub-watershed as spatial units of analysis and service-sheds. For each scenario, soil erosion per unit area in hectares was evaluated at sub-watershed level, and the percentage change was calculated with respect to the baseline conditions, defined by the existing land use and the respective biophysical parameters. This percentage change information is the same used in the last stage of the strategic component, to assess the performance of Watershed Investments (see description in Sect. 3.4.3). Ideally, this is the closing of a first round of the dynamic process of negotiation among stakeholders involved in knowledge use and production for designing and assessing impacts of Watershed Investments.

3.6 Concluding Remarks

This chapter presented an operational approach for designing and assessing impacts of Watershed Investments, building on the concepts of ecosystem services and boundary work. The approach can effectively support the implementation of Watershed Investments that aim at achieving urban water security along with other social or environmental objectives. By addressing stakeholders' concerns of *credibility*, *saliency* and *legitimacy*, it can facilitate negotiation of objectives, definition of scenarios, assessment of alternative Watershed Investments and planning a follow-up, to tackle local challenges, ultimately, to help achieve resilient socio-ecological system. Specifically, the strategic component identifies as key inputs of the process of Watershed Investment design and assessment, the definition of objectives and visioning of feasible and desirable scenarios by stakeholders. The technical component applies spatially explicit modelling to design investments, hence to model the impacts on selected ecosystem services. Ultimately, by addressing the challenges of linking diverse stakeholders and knowledge system across management levels and institutional boundaries, the proposed approach for designing and assessing impacts of Watershed Investments can contribute to implementing adaptive management in the urban water sector, in real-life .

In the remainder of the book, we present a practical application of the proposed approach in a case study of the urban water sector, in Sub-Saharan Africa context. More specifically, Chap. 4 briefly introduces the case of Asmara city and its main water supply, the Toker Watershed, highlighting two key socio-ecological challenges relating to soil erosion and water scarcity. It also presents a central actor in the case study, the Asmara Water Supply Department. Following, Chap. 5 presents an application of the approach for designing and assessing impact of Watershed Investments.

Assuming urban water security and rural poverty alleviation as two objectives for investment, the case study application explores all the steps of the proposed approach.

References

Abson DJ, von Wehrden H, Baumgärtner S, Fischer J, Hanspach J, Härdtle W, Heinrichs H, Klein AM, Lang DJ, Martens P, Walmsley D (2014) Ecosystem services as a boundary object for sustainability. Ecol Econ 103:29–37. https://doi.org/10.1016/j.ecolecon.2014.04.012

Adem Esmail B, Geneletti D (2017) Design and impact assessment of watershed investments: an approach based on ecosystem services and boundary work. Environ Impact Assess Rev 62:1–13. https://doi.org/10.1016/j.eiar.2016.08.001

Adem Esmail B, Geneletti D, Albert C (2017) Boundary work for implementing adaptive management: a water sector application. Sci Total Environ 593–594:274–285. https://doi.org/10.1016/j.scitotenv.2017.03.121

Berkes F (2009) Evolution of co-management: role of knowledge generation, bridging organizations and social learning. J Environ Manag 90:1692–1702. https://doi.org/10.1016/j.jenvman.2008.12.001

Burkhard B, Maes J (eds) (2017) Mapping ecosystem services. Pensoft Publishers. https://doi.org/10.3897/ab.e12837

Cash DW, Clark WC, Alcock F, Dickson NM, Eckley N, Guston DH, Jager J, Mitchell RB (2003) Knowledge systems for sustainable development. Proc Natl Acad Sci 100:8086–8091. https://doi.org/10.1073/pnas.1231332100

Clark WC, Tomich TP, van Noordwijk M, Guston D, Catacutan D, Dickson NM, McNie E (2016) Boundary work for sustainable development: natural resource management at the consultative group on international agricultural research (CGIAR). Proc Natl Acad Sci 113:4615–4622. https://doi.org/10.1073/pnas.0900231108

Crona BI, Parker JN (2012) Learning in support of governance: theories, methods, and a framework to assess how bridging organizations contribute to adaptive resource governance. Ecol Soc 17:32. https://doi.org/10.5751/ES-04534-170132

Daw TM, Brown K, Rosendo S, Pomeroy R (2011) Applying the ecosystem services concept to poverty alleviation: the need to disaggregate human well-being. Environ Conserv 38:370–379. https://doi.org/10.1017/S0376892911000506

de Groot RS, Alkemade R, Braat L, Hein L, Willemen L (2010) Challenges in integrating the concept of ecosystem services and values in landscape planning, management and decision making. Ecol Complex 7:260–272. https://doi.org/10.1016/j.ecocom.2009.10.006

Geneletti D (2013) Assessing the impact of alternative land-use zoning policies on future ecosystem services. Environ Impact Assess Rev 40:25–35. https://doi.org/10.1016/j.eiar.2012.12.003

Geneletti D, Scolozzi R, Adem Esmail B (2018) Assessing ecosystem services and biodiversity tradeoffs across agricultural landscapes in a mountain region. Int J Biodivers Sci Ecosyst Serv Manag 14:188–208. https://doi.org/10.1080/21513732.2018.1526214

Geneletti D (2015) A conceptual approach to promote the integration of ecosystem services in strategic environmental assessment. J Environ Assess Policy Manag 17:1550035. https://doi.org/10.1142/S1464333215500350

Geneletti D, Zardo L, Cortinovis C (2016) Nature-based solutions for climate adaptation: case studies in impact assessment for urban planning. In: Geneletti D (ed) Handbook on biodiversity and ecosystem services in impact assessment. Edward Elgar

Gieryn TF (1983) Boundary-work and the demarcation of science from non-science: strains and interests in professional ideologies of scientists. Am Sociol Rev 48:781. https://doi.org/10.2307/2095325

GIZ, ICLEI (2014) Operationalizing the Urban NEXUS: towards resource-efficient and integrated cities and metropolitan regions. Eshborn

Guerry AD, Polasky S, Lubchenco J, Chaplin-Kramer R, Daily GC, Griffin R, Ruckelshaus M, Bateman IJ, Duraiappah A, Elmqvist T, Feldman MW, Folke C, Hoekstra J, Kareiva PM, Keeler BL, Li S, McKenzie E, Ouyang Z, Reyers B, Ricketts TH, Rockström J, Tallis H, Vira B (2015) Natural capital and ecosystem services informing decisions: From promise to practice. Proc Natl Acad Sci 112:201503751. https://doi.org/10.1073/pnas.1503751112

Haase D, Frantzeskaki N, Elmqvist T (2014) Ecosystem services in urban landscapes: practical applications and governance implications. Ambio 43:407–412. https://doi.org/10.1007/s13280-014-0503-1

Hamel P, Bremer LL, Ponette-González AG, Acosta E, Fisher JRB, Steele B, Cavassani AT, Klemz C, Blainski E, Brauman KA (2020) The value of hydrologic information for watershed management programs: the case of Camboriú, Brazil. Sci. Total Environ 135871. https://doi.org/10.1016/j.scitotenv.2019.135871

Higgins J, Zimmerling A (2013) A primer for monitoring water funds. Global Freshwater Program. The Nature Conservancy, Arlington, VA

Howe C, Suich H, Vira B, Mace GM (2014) Creating win-wins from trade-offs? Ecosystem services for human well-being: a meta-analysis of ecosystem service trade-offs and synergies in the real world. Glob Environ Chang 28:263–275. https://doi.org/10.1016/j.gloenvcha.2014.07.005

Kahneman D, Tversky A (1984) Choices, values, and frames. Am Psychol 39:341–350. https://doi.org/10.1037/0003-066X.39.4.341

Kroeger T, Klemz C, Boucher T, Fisher JRB, Acosta E, Cavassani AT, Dennedy-Frank PJ, Garbossa L, Blainski E, Santos RC, Giberti S, Petry P, Shemie D, Dacol K (2019) Returns on investment in watershed conservation: application of a best practices analytical framework to the Rio Camboriú Water Producer Program, Santa Catarina. Brazil Sci Total Environ 657:1368–1381. https://doi.org/10.1016/j.scitotenv.2018.12.116

Lamorgese L, Geneletti D (2015) Equity in sustainability assessment: a conceptual framework. In: Handbook of sustainability assessment. Edward Elgar Publishing, pp 57–76. https://doi.org/10.4337/9781783471379.00009

Landsberg F, Treweek J, Stickler MM, Henninger N, Venn O (2013) Weaving ecosystem services into impact assessment. a step-by-step method (Version 1.0), 1.0. ed. World Resources Institute, Washington, DC

Lawler JJ, Lewis DJ, Nelson E, Plantinga AJ, Polasky S, Withey JC, Helmers DP, Martinuzzi S, Pennington D, Radeloff VC (2014) Projected land-use change impacts on ecosystem services in the United States. Proc Natl Acad Sci 111:7492–7497. https://doi.org/10.1073/pnas.1405557111

Mandle L, Bryant BP, Ruckelshaus M, Geneletti D, Kiesecker JM, Pfaff A (2016) Entry points for considering ecosystem services within infrastructure planning: how to integrate conservation with development in order to aid them both. Conserv Lett 9:221–227. https://doi.org/10.1111/conl.12201

McDonald RI, Shemie D (2014) Urban water blueprint: mapping conservation solutions to the global water challenge. TNC, Washington, DC

Morrison-Saunders A, Marshall R, Arts J (2007) EIA Follow-Up International Best Practice Principles. Special Publication Series No. 6. Int. Assoc. Impact Assess

Olsson P, Folke C, Galaz V, Hahn T, Schultz L (2007) Enhancing the fit through adaptive co-management: creating and maintaining bridging functions for matching scales in the Kristianstads Vattenrike Biosphere Reserve. Sweden Ecol Soc 12. https://doi.org/28

Pagella TF, Sinclair FL (2014) Development and use of a typology of mapping tools to assess their fitness for supporting management of ecosystem service provision. Landsc Ecol 29:383–399. https://doi.org/10.1007/s10980-013-9983-9

Parker J, Crona B (2012) On being all things to all people: boundary organizations and the contemporary research university. Soc Stud Sci 42:262–289. https://doi.org/10.1177/0306312711435833

Polasky S, Nelson E, Camm J, Csuti B, Fackler P, Lonsdorf E, Montgomery C, White D, Arthur J, Garber-Yonts B, Haight R, Kagan J, Starfield A, Tobalske C (2008) Where to put things? spatial land management to sustain biodiversity and economic returns. Biol Conserv 141:1505–1524. https://doi.org/10.1016/j.biocon.2008.03.022

Ruckelshaus M, McKenzie E, Tallis H, Guerry A, Daily GC, Kareiva P, Polasky S, Ricketts T, Bhagabati N, Wood SA, Bernhardt JR (2015) Notes from the field: lessons learned from using ecosystem service approaches to inform real-world decisions. Ecol Econ 115:11–21. https://doi.org/10.1016/j.ecolecon.2013.07.009

Sharp ER, Chaplin-Kramer R, Wood S, Guerry A, Tallis H, Ricketts T, Authors C, Nelson E, Ennaanay D, Wolny S, Olwero N, Vigerstol K, Penning D, Mendoza G, Aukema J, Foster J, Forrest J, Cameron D, Arkema K, Lonsdorf E, Kennedy C, Verutes G, Kim C, Guannel G, Papenfus M, Toft J, Mar M, Bernhardt J, Griffin R, Glowinski K, Chaumont N, Perelman A, Lacayo M, Hamel P, Vogl AL, Rogers L, Bierbower W, Sharp C, Mandle M, User WI, Natural T, Project C (2015) InVEST 3.2.0 User's Guide. The Natural Capital Project

Star SL, Griesemer JR (1989) Institutional ecology, 'Translations' and boundary objects: amateurs and professionals in Berkeley's Museum of vertebrate zoology, 1907–39. Soc Stud Sci 19:387–420. https://doi.org/10.1177/030631289019003001

Tallis H, Kennedy CM, Ruckelshaus M, Goldstein J, Kiesecker JM (2015) Mitigation for one & all: an integrated framework for mitigation of development impacts on biodiversity and ecosystem services. Environ Impact Assess Rev 55:21–34. https://doi.org/10.1016/j.eiar.2015.06.005

Turner RK, Daily GC (2008) The ecosystem services framework and natural capital conservation. Environ Resour Econ 39:25–35. https://doi.org/10.1007/s10640-007-9176-6

Vogl A, Tallis H, Douglass J, Sharp R, Wolny S, Veiga F, Benitez S, León J, Game E, Petry P, Guimerães J, Lozano JS (2015) Resource investment optimization system (RIOS). In: Introduction and theoretical documentation. Stanford, CA

Chapter 4
Challenges for Water Security
in Asmara, Eritrea

Abstract This chapter presents a case study of urban water sector selected to apply a novel approach for designing and assessing impacts of Watershed Investments and more generally to explore the challenges of urban water security in a Sub-Saharan Africa context characterized by limited resources. The case is about the city of Asmara—the city capital of Eritrea—and its main water supply, the Toker watershed. The chapter illustrates the main socio-ecological challenges and opportunities for promoting adaptive management in the case study. In particular, to illustrate contextual and contingent factors characterizing the case study, three examples of ongoing soil and water conservation activities are presented. The chapter concludes with a focus on the Asmara Water Supply Department (AWSD), a key stakeholder in the selected case study. Specifically, the AWSD is analysed through an approach for conceptualizing water utilities as "learning organization" and assessing their institutional capacity.

Keywords Ecosystem services · Integrated urban water management · Adaptive urban water management · Boundary work · Soil erosion control · Learning organization · Institutional capacity · Water utility maturity model · Water security · Sub-Saharan Africa · UNESCO World Heritage Site

4.1 Introduction

This chapter presents a case study of urban water sector selected to illustrate the application of the approach developed in the previous chapter, and more generally to explore the challenges of water security in a Sub-Saharan Africa context characterized by limited resources. The case study is about Asmara, the capital city of Eritrea, a country of less than 5.2 million inhabitants, located in Eastern Africa.

With almost 60% of its population living in rural areas, Eritrea is one of the last countries facing rapid urbanization (see also Fig. 4.1). According to the 2018 World Urbanization Prospects, during 1988–2018, the urban population had grown from 528,888 (17.6% of the total population) to 2,079,314 (40.1%), mainly in the capital

© The Author(s), under exclusive license to Springer Nature Switzerland AG 2020   39
B. Adem Esmail and D. Geneletti, *Ecosystem Services for Urban Water Security*,
SpringerBriefs in Geography, https://doi.org/10.1007/978-3-030-45666-5_4

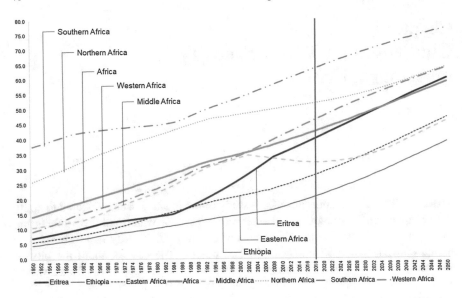

Fig. 4.1 Annual percentage of population residing in urban areas in the study country and region, 1950–2050 (*Source* of data: United Nations, DESA, Population Division (2018))

Asmara. It thus offers an interesting perspective to explore the current challenges of urban water security in the context of a rapidly developing city in Sub-Saharan Africa.

A small village in the Eritrean highlands until the end of the nineteenth century, Asmara started growing after its declaration as the capital city of the new Italian colony in 1900. The city had grown rapidly, especially during the second half of the colonial occupation (1922–41), hosting a sizeable community of Italian settlers: according to the 1939 census, 53,000 of the 98,000 inhabitants of Asmara were Italians. Moreover, the city acquired a distinct Italian architectural look to the extent of earning the appellative of *Piccola Roma* (Little Rome) and several decades later, in 2017, it was declared as a UNESCO World Heritage Site for its well-preserved modernist architecture.

Today, with roughly half million inhabitants, Asmara is the only major city in Eritrea, accounting for almost 10% the total population and half of urban population in the country. Figure 4.2 presents some illustrative panoramic views of the city, highlighting differences between neighbourhoods characterized by different levels of access to services and generally socio-economic conditions of their residents. Further population growth will have great implications in terms of water supply and consumption, housing and for the provision of physical and social infrastructure, as well as goods and services. Particularly, water supply is considered the limiting factor for population growth and city development, as described in the following Sect. 4.2, while the remainder of this chapter illustrates key challenges associated with water security in Asmara, describes opportunities from ongoing watershed initiatives and illustrates the main features of the water utilities.

Fig. 4.2 Panoramic view of Asmara facing South-East direction (top, By Hansueli Krapf—Own work, CC BY-SA 3.0); areal view of a well-off neighbourhood (left, By John Beso, CC BY-SA 3.0) and a neighbourhood with tightly packed homes in an old quarter of Asmara (right, By David Stanley, CC BY 2.0). Panoramic views of Asmara: (top) view of the city facing South-East direction, photo by Hansueli Krapf - Own work, CC BY-SA 3.0, https://commons.wikimedia.org/w/index.php?curid=7106467, (bottom-left) a well-off neighborhood, facing northeast, photo by John Beso, CC BY-SA 3.0, https://commons.wikimedia.org/w/index.php?curid=46699092, (bottom-right) a neighborhood with tightly packed homes in an old quarter of Asmara, photo by David Stanley from Nanaimo, Canada, CC BY 2.0, https://commons.wikimedia.org/w/index.php?curid=24199908

4.2 Water Supply

The main source of water for Asmara is surface water collected as runoff during the rainy seasons. When Asmara was first established, the main source consisted of about nine shallow wells in *Sembel* neighbourhood (southwester part of Asmara), and the water from these wells was treated in the *Godaif treatment* plant (southern part). Later on, however, the wells became unusable as source due to the deterioration of their water quality. The other sources were represented by two lakes in the neighbourhood of *Akria* (north of Asmara), and the related treatment plant. These also had to be stopped from supplying the treatment plant because to the deterioration of the water quality caused by the human settlements around the lakes. Therefore, to meet the increasing water demand of the rapidly growing city, gradually, several micro and medium-sized dams had been constructed on the outskirts of Asmara, including *Maisirwa, Adi-Sciaca, Quazien, Beleza, Adi-Nifas* and *Valle-Gnecchi*. Between 1968 and 1972, the biggest raw water reservoir of *Mai Nefhi* was realized with a nominal capacity of 26 million m^3. Again, to meet the increasing water demand of the rapidly

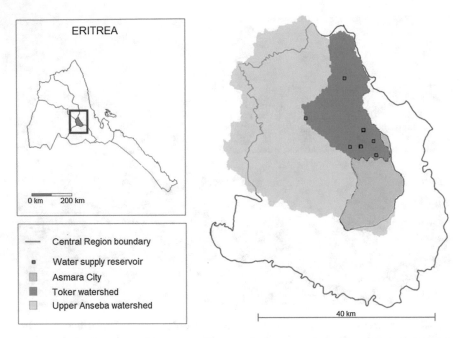

Fig. 4.3 Map of Eritrea and it six administrative regions based on the main watersheds; highlight of the Central Region, the smallest and most densely inhabited region (left). Zoom into the Toker watershed, a sub-watershed of the Upper Anseba in the Central Region, and the seven reservoirs supplying water to Asmara (right) (*Source* Adem Esmail and Geneletti 2017)

growing new capital city of Eritrea, the Toker Dam was completed in 2000, providing an additional supply of about 14,000 m^3 per day of treated water to the existing production of 16,500 m^3 per day.

With an estimated value of US$44 million, the Toker reservoir is indeed one of the most strategic water infrastructures serving Asmara and its surroundings. It thus offers an insightful case study to explore challenges and opportunities for water security in the Sub-Saharan Africa context. In particular, we focus on the Toker watershed, a sub-watershed of the Upper Anseba, considering that it represents the main water supplier for Asmara. As shown in Fig. 4.3, both Asmara and the Toker Watershed are located in the central region or *Zoba Maekel*. This is the smallest and most densely inhibited region in Eritrea, covering less than 1.2% of the total area yet hosting almost 17% of the total population. Noteworthy, the central region is one of the six administrative regions based on the main watersheds in the country, introduced in 1996 after the declaration of the formal independence in 1993, replacing the previous division into nine regions introduced during the colonial era.

4.3   Soil Erosion and Water Scarcity

Two among the most critical socio-ecological challenges in Asmara and the Toker Watershed are soil erosion and water scarcity. Soil erosion is primarily attributed to a long history of poor cultivation and overgrazing, unregulated wood and timber harvesting, lack of recycling of nutrients and poor management of organic matter, as well as rapid urbanization and demographic growth in the case study (Murtaza 1998; Tewolde and Cabral 2011). Water scarcity, on the other hand, is mainly due to persistent droughts that have been associated also with climate variability and change (Abraham et al. 2009; IPCC 2014; MoLWE 2012).

As illustrated in the diagram in Fig. 4.4, the concept of ecosystem services offers an effective way of framing the socio-ecological challenges in Asmara and the Toker Watershed. With respect to soil erosion, for example, the concept can be useful to highlight the spatial mismatch between areas ecosystem services production (i.e.

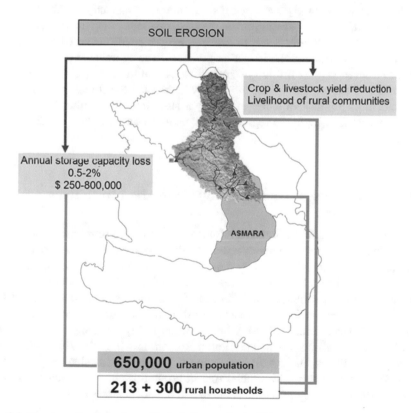

Fig. 4.4 Framing of soil erosion-related challenges in the Toker Watershed from an ecosystem services perspective highlighting (i) spatial mismatch between ecosystem services production and benefit areas, (ii) different impacts on urban and rural beneficiaries

the watershed) and benefit (i.e. the urban areas). Hence, it helps explore how these socio-ecological challenges diversely affect different groups of people (e.g. Daw et al. 2011). Soil erosion causes a rapid loss of storage capacity of reservoirs supplying the city of Asmara: Abraham et al. (2009) have estimated an average sediment yield in the region of 856 t/Km^2, corresponding to an annual storage capacity loss between 0.5 and 2%. In monetary terms, considering the estimated value of $44 million of the Toker reservoir, this would amount to an annual loss of between $250 and $800,000. At the same time, soil erosion affects livelihood of rural communities by resulting in lower yields. A study by the Food and Agriculture Organization (FAO) has estimated that a rate of soil erosion of 1500 t/Km^2 per year could reduce yields by 0.2–0.4% a year for crops and 0.05–0.1% for livestock (see Habtetsion and Tsighe 2007).

Concerning water scarcity, as reported in Box 4.1, over the years, several reservoirs have been built to store surface water, during two wet seasons known as *kiremti* (June–September) and *asmera* (March–April). These reservoirs represent the main sources of water for meeting urban and rural demands, including irrigation, livestock watering, domestic water supply and other uses. According to a study by Abraham et al. (2009), the total number of reservoirs in the Upper Anseba Watershed is 49, of which the 11 biggest ones supply water to Asmara, and 38 smaller reservoirs serve rural communities for drinking and irrigation purposes. The aggregated storage capacity of the 49 reservoirs is 32 million m^3, of which 24.8 million m^3 (77.4%) is reserved for Asmara. However, soil erosion is rapidly decreasing the storage capacity of the reservoirs (only 55–89% of storage capacity is still available), further compounding physical water scarcity in the region with economic water scarcity (Abraham et al. 2009).

4.4 Opportunities from Ongoing Watershed Initiatives

There are several ongoing watershed initiatives, which could represent valuable windows of opportunity for implementing an adaptive management. By way of example, we selected three initiatives, briefly described hereafter and more details in Boxes 4.1, 4.2 and 4.3. The selected initiatives are illustrative of the contextual and contingent factors as well as the relative influence of stakeholders in the study area. A first and most significant initiative consists of existing partnerships between the Asmara Water Supply Department (AWSD), i.e. the water utility that supplies water to the city of Asmara, and farmers in the Central Region, in which the Toker Watershed is located (see Box 4.1). As will be illustrated later on in this chapter, the AWSD is indeed a central actor with crucial role in promoting the implementation of an adaptive management paradigm in Asmara and the Toker watershed.

> **Box 4.1: First illustrative initiative in the Toker Watershed. Existing partnership between utility and farmers in the Central Region (Source Abraham et al. 2009)**
>
> Asmara Water Supply Department (AWSD) provides agricultural extension services to farmers at sub-zoba branch offices. There are five associations in Central Zone (Zoba Maekel), each having a management committee, consisting of a chairperson, secretary and treasurer. The associations include 1.126 farmer members engaged in: horticultural production, cattle fattening, beekeeping, poultry, and dairy production. Among other things, the "Toker Project" provides technical and financial support to farmers in the watershed, for instance, by running 10 village shops to ease access to agricultural inputs (fertilizers, chemicals…). The initiative is coordinated by a management committee that consist of representatives of farmers, village administration and the project (NFIS 2005 as cited by Abraham et al 2009).

A second initiative is a transdisciplinary research project dealing with water resource management in the Upper Anseba Watershed of which the Toker is a sub-watershed (see Fig. 4.3). Among others, the projects developed a sound spatially explicit database, including the position and status of reservoirs, beneficiaries and relevant biophysical data, which had been used in this research. However, it did not explicitly explore potential urban-rural partnerships. Perhaps one limitation is the fact that it is more concerned with the rural implications of water resources management, overlooking the urban-rural interactions (see Box 4.2).

> **Box 4.2: Second illustrative initiative in the Toker Watershed. Transdisciplinary research on water resource management in the Central region, focusing on agricultural use (Source: Abraham et al. 2009)**
>
> **About the program**
>
> The "Appraisal of Surface Water in the Upper Anseba Watershed" (ASW-UAW) is a transdisciplinary research aiming to create a basis for informed decision-making processes in the use of surface waters in the Central Region, and more specifically the Upper Anseba Watershed. It addressed key shortcomings in: (i) the information required for an efficient management of surface waters, and (ii) the participation of stakeholders.
>
> Its objective was to assess surface water capacity and management, raise awareness, and build capacity of the major stakeholders. More specifically, to:
>
> A. Create a spatial database, high-resolution satellite image maps to address the shortcomings in the information required as a basis for informed decisions for more efficient management of surface waters, i.e., to fair allocation of resources according to the needs of the population and balanced with the capacity of the catchment to generate the required water resource.

B. Evaluate the general characteristics and problems of reservoirs in Central Region
    with emphasis in the Upper Anseba Catchment.

C. Assess the extent and efficiency of water use with a focus on the irrigation system
    and estimate the potential irrigable areas.

D. Estimate the extent of sediment deposition of selected reservoirs.

E. Assess community perceptions & ambitions regarding the reservoirs & their use.

F. Identify promising practices, methodologies and approaches that can be a
    basis for replication in other catchments as pilot for similar studies and
    implementations.

*Note: The ASW-UAW was funded by the Eastern and Southern Africa Partnership
Programme and supported by the Swiss Centre for Development and Environment,
within the framework of the Sustainable Land Management Programme, Eritrea.*

A third initiative is a so-called *Summer Student Work Program* (SSWP). Launched
by the Ministry of Education (MoE) in 1994, the SSWP engages secondary school
students in a wide range of activities, including forestation, soil and water conservation, and assisting poor farmers. It is a valuable socio-ecological "experiment",
allowing students with urban background to reconnect to nature and interact with
farmers of different social-ethnic-economic extraction. At the same time, it con-
tributes to the restoration of ecosystem and their services, often directly benefiting
the rural communities, as well as assists poor farmers. Of interest is an assessment of
the first 15 years of the SSWP, carried out in 2009, which is a milestone of the "social
learning" around soil and water conservation and generally watershed management
in Eritrea (see Box 4.3).

**Box 4.3: Third illustrative initiative in the Toker Watershed. Summer
Student Work Program and the involvement of rural communities
(Source: Assessment report by the Eritrean Ministry of Education)**

About the program

Launched by the Ministry of Education (MoE) in 1994, the Student Summer Work
Program (SSWP) engages secondary school students in a wide range of activities,
including forestation, soil and water conservation, and assisting poor farmers. During
1994–2008, the SSWP had a total cost of US$11 million and took place in 182 location
all over the country, of which 17 in our case study region, i.e. Central Region (Zoba
Maekel).

From the perspective of an adaptive watershed management, particularly interesting
is a comprehensive assessment of the SSWP carried out by the MoE in 2009. Its aim
was to assess the level of success of the SSWP, evaluate the perception of students,
teachers and villagers and assess the organization and management of the SSWP.
It considered 62 out of 187 locations of the campaign, involving 400 students, 400
teachers, 186 villagers and various experts.

Involvement of rural communities

The assessment identified the key criteria used for selecting the sites of intervention as well as analysed how they relate to the level of success of the SSWP. The selection criteria included top soil depth and type, slope gradient, and management type. Most interestingly, the assessment highlighted a clear mismatch between participation of farmers in site selection, and actual activities of the SSWP (see graph below). In 53, 36, and 12 percent of the site participation of farmers in site selection was respectively, high, medium, and low. For participation on actual activities, on the other hand, the opposite trend was observed. Moreover, from the interviews it emerged that farmers were not satisfied with the work done by the students, whom they perceive as being too "urban". Instead, they argue they could have achieved better results with the same resources of the program, which is a particularly relevant for exploring the willingness of rural communities to take active part in WI related activities.

Participation of the rural communities in site selection and S&W conservation activities.

Arguably, these three initiatives represent valuable windows of opportunity to promote an adaptive watershed management. To start with, singularly, the three initiatives could benefit from relevant ecosystem services knowledge. For the first initiative, the partnership between the AWSD and the farmers, introducing an ecosystem services perspective has the potential to boost the existing cooperation (e.g. by introducing Payment for ecosystem services schemes). For the second initiative, i.e. the transdisciplinary research on water resources, an ecosystem services approach could provide an important support by addressing the urban-rural linkages, thus overcoming its main shortcoming. Finally, for the SSWP, an ecosystem service approach could ensure that its activities are designed based on sound scientific information. Whereas currently the identification of the areas of intervention heavily relies on expert-based approaches, which lack the needed flexibility to form the basis of an iterative science-informed decision support system.

More in general, what emerges is a need for operational approaches to coordinate such ongoing activities in the watershed. Building on innovative concepts like ecosystem services and boundary work, such approaches have the potential to trigger a process of social learning involving local stakeholders, within a framework of adaptive watershed management. In this, the AWSD can be envisaged a central actor of the process; however, a question that arises is if it has the needed institutional capacity?

4.5 Role of the Water Utility

Indeed, water utilities are recognized as a central actor for operating and maintaining the urban water sector (e.g. Lieberherr and Truffer 2015). Different studies have explored how to better characterize their role as "learning organizations" in the implementation of an adaptive management. A learning organization is here defined as "an organization that is skilled at creating and acquiring knowledge and modifying its behaviour to reflect new insights" (Cowling et al. 2008). In particular, we here refer to the work Oby Kayaga et al. (2013) who have explored the notion of water utilities as learning organization, and proposed their so-called "Water Utility Maturity model" (WUM). Described hereafter, the WUM consists of a Otool for evaluating the institutional capacity of water utilities. It can serve to characterize a given water utility, locating it along a spectrum of institutional capacity, as well as to identify the determinants of capacity that are the most significant for understanding the role of water utilities in implementing adaptive management. Following is a brief illustration of the WUM and its application to characterize the Asmara Water Supply Department.

4.5.1 Water Utility Maturity Model (WUM)

Developed under the auspices of the World Bank, the WUM is a conceptual framework for evaluating institutional capacity of water utilities developed by Kayaga and colleagues. It is a tool with a sound theoretical basis, because it relies on an in-depth analysis of existing conceptualizations of institutions. At the same time, it is oriented to real-life application: it builds on a comparative analysis of numerous existing tools for evaluating institutional capacity (Cullivan et al. 1988; Baietti et al. 2006; Locussol and van Ginneken 2008; Suez Environment 2010, as cited in Kayaga), of the water sector (Saleth and Dinah 2004; AMCOW et al. 2006, 2011; Gandhi et al. 2009, as cited in Kayaga) and of generic international development interventions (Lusthaus et al. 1995; DFID 2003; EU Commission 2009; Kimata 2008, as cited in Kayaga).

A key aspect of the WUM is its emphasis on the concept of learning organizations. For instance, institutional capacity is defined as the capacity of an organization to "continuously generate a minimum level and quality of valued outputs, and to prioritize learning for continuous improvement" (Kayaga et al. 2013). Whereas institutions, i.e. the "rules" and "roles" by which decision-making and implementation is structured, are conceptualized as a "combination of organizations, institutional mechanisms and institutional orientations" (Kayaga et al. 2013). Organizations being the most "tangible" class of institutions that structure the choice of action of individual or corporate and other collective actors within a society. While mechanisms and orientations represent, respectively, the explicit (or formal) and the implicit (or informal) systems of rules. Therefore, in the WUM model, water utilities are conceptualized as "organizational institutions (actors), which operate under, and are constrained by, the overall legal and institutional environment (rules)".

In terms of institutional capacity, the WUM identifies five core capabilities that enable a water utility to "perform and survive in a turbulent operating environment". In particular, it follows an approach proposed by "The European Centre for Development Policy Management", which emphasizes the role of endogenous factors in determining the capacity of institutions, rather than external factors (e.g. foreign expertise). The five core capabilities are, namely, the capabilities to commit and engage; to carry out technical, service delivery and logistical tasks; to relate and attract resource and support; to adapt and self-renew and to balance coherence and diversity (Kayaga et al. 2013). As shown in Fig. 4.5, in the case of WUM, the five capacity dimensions were labelled as *Behaviour*; *Structure/processes*; *Capabilities*; *Organizational tools* and *Influence*. In the context of this chapter, what is the most important is that the five institutional capacity dimensions are "integrative, mutually exclusive and collectively exhaustive". Each capacity dimension has four to five

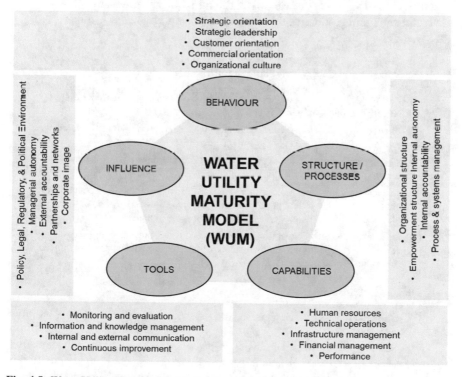

Fig. 4.5 Water Utility Maturity Model, a conceptual framework and related operative tools for evaluating institutional capacity of a water utility (*Source* Kayaga et al. 2013)

attributes (23 attributes in total) that are characterized by five progressive levels of institutional capacity referred to as "maturity levels", namely, (1) initial, (2) basic, (3) proactive, (4) flexible and (5) progressive.

By way of example, Box 4.4 presents the five attributes of the institutional capacity dimension *Influence* and their respective five progressive maturity levels. This specific capacity dimension, which relates to the water utility's "ability to influence its operating environment in a positive and strategic manner" (Kayaga et al. 2013), is indeed the most significant dimension to explore within the scope of this chapter. More specifically, Fig. 4.6 shows an example of the labelled progressive levels of maturity of the attribute *Partnerships and networks*. This is one of the five attributes of the capacity dimension Influence. Worth of notice is how "learning" takes place overtime (x-axis), suggesting the WUM is both a diagnostic (as in the case study application in this chapter) as well as a benchmarking tool. It can be used to identify both barriers to progressing between maturity levels and potential enablers to overcome such barriers.

An empirical application of the WUM, considering the case of the Asmara Water Supply Department, is presented in the following section.

Box 4.4: Five attributes of the institutional capacity dimension "Influence" and respective five progressive maturity levels (Source: Kayaga et al. 2013)

Policy, Legal, Regulatory, and Political Environment

(1) Leadership and staff not well conversant with factors in the external environment. Negative political influence is common.

(2) Leadership passively interested in factors in the external environment, and reacts to them rather than strategically influencing them.

(3) The external environment is actively monitored to develop understanding & reduce uncertainty.

(4) Leadership continuously scanning the external environment, and adapting to changes through building organizational capacity for effective negotiation, and alignment of business processes, building networks and allies.

(5) Utility has predictive capabilities, and carries out risk/opportunities assessment & and management; continuously adaptive to the external environment in near real-time.

Managerial autonomy

(1) Utility managers lack autonomy to make important managerial and operational decisions.

(2) There is limited managerial and operational autonomy.

(3) Managers have more room to manoeuvre and innovate (i.e. have autonomy to effect internal managerial/operational changes to improve the effectiveness and productivity).

(4) Utility has full autonomy with respect to most managerial, operational and financial decisions.

(5) Utility has full autonomy with respect to all managerial, operational and financial decisions.

External accountability

(1) There is no external accountability for performance.

(2) External accountability mechanisms in place but not effective.

(3) The utility is held accountable for performance by some of the external stakeholders.

(4) Utility is held accountable for performance by some external stakeholders.

(5) Utility has a balanced accountability framework.

Partnerships and networks

(1) Partnerships and networks with outside organizations are not supported.

(2) Partnerships and networks may be initiated by individual staff. Supplier communications are limited to tendering, order placement or problem resolution.

(3) There is a policy that encourages and supports mutually beneficial partnerships and networking. Processes are in place to select, evaluate and rank suppliers.

(4) There is a budge to develop and grow partnerships and networks. Relationship processes exist to develop key suppliers.

(5) Partnerships are integrated within business processes.

Corporate image*

(1) Corporate image is not recognized as an important service element and is not evaluated.

(2) Leadership is aware of the importance of corporate image; however, it is not monitored or evaluated in a consistent and systematic manner.

(3) Corporate image is periodically measured; but the results are not necessarily used for improvements.

(4) Corporate image is continuously and systematically tracked. The results are widely made available inside the organization and used in the strategic planning process.

(5) The results of the corporate image scans are integrated into the performance/incentive management system for staff.

4.5.2 Asmara Water Supply Department

A central actor in the study area, the Asmara Water Supply Department (AWSD) is a public utility managing water and sanitation services, serving 350,000 people out of a target population of 450,000 in Asmara and surrounding districts, i.e. a coverage of 77% (Zeraebruk et al. 2014). The AWSD serves 34,128 connections with 7.46 million m^3 of water delivered, yearly. It employees 460 people (i.e. 13.48 every 1,000 users) and has an annual sales revenue of US$ 4.85 million. As for its water resources, the AWSD mainly relies on surface water collected as runoff during two wet seasons (*kiremti* and *asmera*), stored mainly in four reservoirs. These include the Toker, Adisheka and Maisirwa reservoirs, which are located to the north of Asmara, and the Mai Nefhi dam, located southwest of Asmara, in a sub-watershed of the Barka River. Finally, the AWSD operates three waterworks, i.e. *Stretta Vaudetto*, *Adinfas* and *Mai Nefhi*, amounting to a production capacity of 44,000 m^3 per day, often reduced to half due to technical problems, aging infrastructure and at times due to limited volume of water in reservoirs (Fig. 4.6).

Applying the WUM by (Kayaga et al. 2013), the AWSD has been evaluated in terms of its institutional capacity. Operationally, the evaluation was based on information from reports, the literature and an interview with a key informant. Overall, the AWSD resulted in being a *Level-2 water utility with Basic maturity*. The assigned level refers to all the 23 attributes of the five capacity dimensions of the WUM. By way of example, Table 4.1 presents the characterization of the AWSD with respect to the capacity dimension *Influence* and its five attributes. Here, the assumption is that the *Influence* dimension is the most significant in terms of implementing adaptive management in the urban water sector. Beyond the specific case study, worth highlighting here is the potential of the WUM to provide a detailed characterization of a water utility and its institutional capacity, as a proxy of its ability to become a

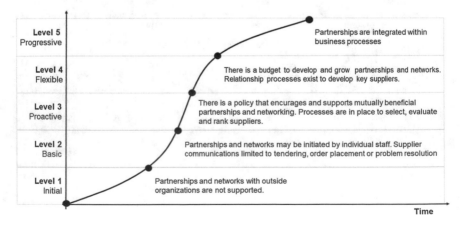

Fig. 4.6 Progressive maturity levels for the attribute "Partnership and networks" of the capacity dimension "Influence" (*Source* Kayaga et al. 2013)

Table 4.1 Characterizing the Asmara water supply department under five attributes of the capacity dimension "Influence" of the water utility maturity model

Attribute	Description of the maturity level: basic level 2
Policy, legal, regulatory and political environment	Leadership passively interested in factors in the external environment, and reacts to them rather than strategically influencing them
Managerial autonomy	There is limited managerial and operational autonomy
External accountability	External accountability mechanisms in place but not effective
Partnerships and networks	Individual staff may initiate partnerships and networks. Supplier communications are limited to tendering, order placement or problem resolution
Corporate image	Leadership is aware of the importance of corporate image; however, fails to monitor or evaluate it in a consistent and systematic manner

central actor in the implementation of an adaptive watershed management. Above all, the WUM is a promising tool also due to its solid theoretical foundation and strong orientation towards real-life application.

4.6 Concluding Remarks

This chapter presented the case of Asmara city and its main water supply, the Toker Watershed, highlighting two key socio ecological challenges. In particular, it highlighted how soil erosion and water scarcity hinder the city of Asmara from meeting its growing water demands. At the same time, they seriously jeopardize the sources of income of the rural communities, whose livelihood depends primarily on rainfed agriculture.

Shifting the focus to the role of actors, this chapter also aimed at gaining a better understanding of the role of water utilities in the urban water sector and, more specifically, of their role as learning organization implementing adaptive management. To this end, it briefly introduced a tool for evaluating institutional capacity of a water utility, the WUM model. The latter is indeed a promising tool with a solid theoretical basis and strong orientation towards real-life application. Following, the WUM model was applied characterize the AWSD as a central actor and learning organization that could potentially drive the implementation of an adaptive management in the case study. Specifically, an evaluation of its institutional capacity was carried out using the WUM and focusing on the attributes of the *influence* capacity dimension.

In the following chapter, two guiding policy goals are assumed dealing with how to secure water for the city while also addressing the issues of poverty affecting the rural communities.

References

Abraham D, Tesfaslasie F, Tesfay S (2009) An appraisal of the current status and potential of surface water in the upper Anseba catchment potential of surface water in the upper Anseba Catchment. https://doi.org/10.7892/boris.69703

Adem Esmail B, Geneletti D (2017) Design and impact assessment of watershed investments: an approach based on ecosystem services and boundary work. Environ Impact Assess Rev 62:1–13. https://doi.org/10.1016/j.eiar.2016.08.001

Cowling RM, Egoh B, Knight AT, O'Farrell PJ, Reyers B, Rouget M, Roux DJ, Welz A, Wilhelm-Rechman A (2008) An operational model for mainstreaming ecosystem services for implementation. Proc Natl Acad Sci 105:9483–9488. https://doi.org/10.1073/pnas.0706559105

Daw TM, Brown K, Rosendo S, Pomeroy R (2011) Applying the ecosystem services concept to poverty alleviation: the need to disaggregate human well-being. Environ Conserv 38:370–379. https://doi.org/10.1017/S0376892911000506

Habtetsion S, Tsighe Z (2007) Energy sector reform in Eritrea: initiatives and implications. J Clean Prod 15:178–189. https://doi.org/10.1016/j.jclepro.2005.09.003

IPCC (2014) Summary for policy makers. In: Field CB, Barros VR, Dokken DJ, KJ, M, Mastrandrea MD, Bilir TE, Chatterjee M, Ebi KL, Estrada YO, Genova RC, Girma B, Kissel ES, Levy AN, MacCracken S, Mastrandrea PR, White LL (eds) Climate change 2014: impacts, adaptation and vulnerability—contributions of the working group II to the fifth assessment report of the intergovernmental panel on climate change. Cambridge University Press, Cambridge, UK, New York, NY, USA, pp 1–32

Kayaga S, Mugabi J, Kingdom W (2013) Evaluating the institutional sustainability of an urban water utility: a conceptual framework and research directions. Util Policy 27:15–27. https://doi.org/10.1016/j.jup.2013.08.001

Lieberherr E, Truffer B (2015) The impact of privatization on sustainability transitions: a comparative analysis of dynamic capabilities in three water utilities. Environ Innov Soc Trans 15:101–122. https://doi.org/10.1016/j.eist.2013.12.002

MoLWE (2012) Eritrea's second national communication under the united nations framework convention on climate change (UNFCCC). Asmara, Eritrea

Murtaza N (1998) The pillage of sustainablility in Eritrea, 1600s–1990s: rural communities and the creeping shadows of hegemony. Greenwood Press, Westport, Connecticut

Tewolde MG, Cabral P (2011) Urban sprawl analysis and modeling in Asmara. Eritrea Remote Sens 3:2148–2165. https://doi.org/10.3390/rs3102148

United Nations, Department of Economic and Social Affairs, Population Division (2018) World Urbanization Prospects: The 2018 Revision, Online Edition

Zeraebruk KN, Mayabi AO, Gathenya JM, Tsige Z (2014) Assessment of level and quality of water supply service delivery for development of decision support tools: case study Asmara water supply. Int J Sci Basic Appl Res 14:93–107. https://doi.org/10.5539/enrr.v4n4p208

Chapter 5
Designing Watershed Investments for Asmara and the Toker Watershed

Abstract This chapter presents an application of a novel operational approach for designing and assessing the impacts of Watershed Investments, developed in Chap. 3, to the Asmara and Toker Watershed case study. Assuming urban water security and rural poverty alleviation as two objectives for Watershed Investments, the case study application explores all the steps of the proposed approach. The results of the application include spatially explicit data that allow quantitatively assessing the performance of different Watershed Investment scenarios in terms of changes in a selected ecosystem service, answering to important planning and management questions. The application to the Asmara and Toker Watershed case study also highlights the challenges of addressing stakeholders' concerns through relevant boundary work strategies.

Keywords Ecosystem services modelling · Integrated urban water management · Adaptive urban water management · Boundary work · Soil erosion control · InVest · Resources investment optimization system RIOS · Natural capital project · Eritrea · Water security · sun-Saharan Africa

5.1 Introduction

This chapter presents the application of the approach presented in Chap. 3 for designing, and assessing impacts of Watershed Investments to the Asmara city and Toker Watershed case study. Considering the challenges associated with soil erosion and water scarcity, illustrated in Chap. 4, urban water security and rural poverty alleviation are chosen as two illustrative objectives for Watershed Investments. Figure 5.1 presents an overview of the case study application.

The focus of the chapter is on the application of the technical component (Sect. 3.5) in a data-scarce context, whereas the strategic component (Sect. 3.4) is mainly based on review of documents and interviews with key informants, and serves to set a real-life socio-ecological background. Ultimately, the chapter aims to highlight the potential of the proposed approach to support a process of stakeholder negotiations for Watershed Investments design and impacts assessment. To this end, it presents some illustrative outputs allowing answering key planning and management questions,

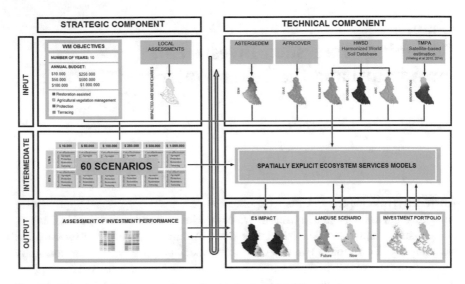

Fig. 5.1 Overview of the case study application in the Toker Watershed

such as (i) *which activities, when and where yield the greatest returns*? and (ii) *what is the impact on a selected ecosystem service?*

The results of the application include spatially explicit data that allow to quantitatively assessing the performance of different investment scenarios in terms of changes in a selected ecosystem service, answering to key planning and management questions. The application also shows how, by addressing stakeholders' concerns of *credibility*, *saliency* and *legitimacy*, the proposed approach can facilitate negotiation of objectives, definition of scenarios and assessment of alternative Watershed Investments.

In the remainder of the chapter, first, the baseline conditions in the Toker Watershed are introduced. Hence, results related to the first two stages of the strategic component: setting the agenda and defining investment scenarios are presented. Following, for the technical component, some illustrative results concerning biophysical data processing; investment portfolios and impact on soil erosion control are reported. Finally, results for the last stage of the strategic component, dealing with assessing the performance of investment scenarios and planning for a follow-up are presented. This includes tables that synthesize the investment portfolios in the case study as well as a table that shows the performances of Watershed Investments at sub-watershed level.

5.2 Baseline Conditions

The existing distribution of land use and land cover (LULC) in the Toker Watershed is illustrated in Fig. 5.2, based on a simplified classification adopted by the RIOS approach (see Table 5.1 and Vogl et al. 2015). The map also shows the reservoirs and related sub-watersheds in the Toker Watershed, which are used as, respectively, entry points and spatial units for our analysis. As mentioned earlier in Chap. 4, the Toker Watershed is the main water supply for Eritrea's capital city, Asmara, located in the most densely inhabited region, the Central Region, and accounting roughly for 10 percent of the total population. Soil erosion and water scarcity are two complex socio-ecological challenges, emerging among the most critical issues that require urgent action (Murtaza 1998; Tewolde and Cabral 2011; Abraham et al. 2009; IPCC 2014; MoLWE 2012).

From a boundary work perspective, reaching consensus over the baseline conditions (e.g. prioritizing socio-ecological challenges, agreeing on the LULC distribution or adopting sub-watershed instead of other administrative boundaries as units of analysis and so on) can represent a good example of the achievement of a boundary object. As such, it requires adequate participation of key stakeholders (e.g.

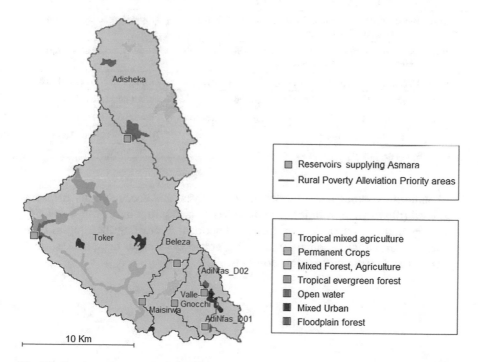

Fig. 5.2 Land use in the Toker watershed and location of the 7 reservoirs and sub-watersheds supplying water to Asmara. Land use classes based on a simplified RIOS classification described in Vogl et al. (2015). Highlight of the rural poverty alleviations priority areas, i.e. AdiSheka (north) and AdiNfas_D01 (south) (*Source* Adem Esmail and Geneletti 2017)

Table 5.1 Reclassification from Africover to RIOS General land use classes

AFRICOVER: landcover database for Eritrea (User defined classes)	RIOS general land use classes
Closed semi-evergreen trees with closed to open shrubs + Rainfed Herbaceous Crop agriculture	Mixed forest, agriculture
Open shrubs with closed to open herbaceous and sparse trees + Rainfed Herbaceous Crop agriculture	Mixed forest, agriculture
Forest plantation—Eucalyptus + Rainfed Herbaceous Crop, Small Fields, Clustered—Cereal	Mixed forest, agriculture
Urban area	Mixed urban
Artificial lake	open water
Irrigated Non-Graminoid Crop	Permanent crops
Sparse shrubs with sparse herbaceous	Shrub/scrub
tropical mixed agriculture	Tropical mixed agriculture
Bare soil + Rainfed Herbaceous Crop	Tropical mixed agriculture
Rainfed Herbaceous Crop, Small Fields + Irrigated Non-Graminoid Crop, Small Fields, Clustered—Vegetables	Tropical mixed agriculture
Rainfed Herbaceous Crop, Small Fields—Cereal + Sparse trees with sparse herbaceous	Tropical mixed agriculture
Rainfed Herbaceous Crop, Small Fields—Cereal + Forest Plantation, Clustered - Eucalyptus	Tropical mixed agriculture
Irrigated Non-Graminoid Crop, Small Fields—Vegetables + Forest Plantation, Clustered - Eucalyptus	Tropical mixed agriculture
Rainfed Herbaceous Crop, Small Fields—Cereal + Bare soil	Tropical mixed agriculture
Forest plantation—Eucalyptus AND Artificial lake	Floodplain forest
Artificial lake AND Forest Plantation, Clustered—Eucalyptus	Floodplain forest
Forest plantation-Eucalyptus	Tropical evergreen forest

Source Vogl el at. (2015)

local authorities, experts, and farmer representatives) and accountability measures (e.g. legal role of water utilities to manage watersheds), to jointly ensure *saliency*, *legitimacy* and in part *credibility*. Equally important is creating meaningful communication among stakeholders, translating concepts to make them accessible to laypersons and mediating possible conflicts.

5.3 Setting the Agenda and Defining Investment Scenarios

5.3.1 Setting the Agenda

5.3.1.1 Defining Investment Objectives and Planning Horizon, and Prioritizing Ecosystem Services

Given the challenges related to soil erosion and water scarcity described in Chap. 4, urban water security and rural poverty alleviation were defined as two illustrative Watershed Investment objectives, with respect to a planning horizon of 10 years. Accordingly, and by way of example, soil erosion control was considered as the most relevant ecosystem service, to be used for the Watershed Investment design and assessment. For the first objective, aiming at decreasing sediment yield to reservoirs, RIOS was applied to design investment portfolios. Likewise, RIOS was applied for the second objective; yet, in this second case, spatial constraints that define investment priority areas were added. These spatial constraints consist of areas of the watershed where activities are to be preferred or avoided irrespective of their cost-effectiveness. In the Toker Watershed case study, activities in the *AdiSheka* and *AdiNifas_D01* sub-watersheds were prioritized (see sub-watershed highlighted in blue in Fig. 5.2), assuming rural communities could benefit from the Watershed Investments. Benefits could be both in terms of poverty reduction (e.g. financial resources that integrate their livelihood in exchange for maintenance of ecosystem services) and poverty prevention (e.g. erosion control that increases crop and livestock yield).

5.3.1.2 Characterizing Stakeholders

In the Toker Watershed case study, given the focus on soil erosion control, watersheds represent the actual service-sheds. Therefore, reservoirs and their sub-watersheds can be used, respectively, as *entry points* for identifying who utilizes the reservoirs and *spatial units* for exploring the linkages between ecosystem services, activities and different groups of stakeholders. Operatively, the Toker Watershed was divided into seven sub-watersheds, corresponding to seven reservoirs that supply the city of Asmara. Based on findings of a previous study, the beneficiaries were divided into urban, i.e. inhabitants of Asmara, and rural, i.e. 213 and 300 rural households, benefiting from *AdiSheka* and *AdiNifas_D01* reservoirs, respectively (Abraham et al. 2009). Moreover, urban and rural beneficiaries were reasonably assumed to be uniformly distributed over the city area and the sub-watershed of interest, respectively. This can be an acceptable approximation, given the focus on the urban-rural divide and not on the differentiated access to benefits within the city itself. Finally, collecting new socio-economic and demographic data was beyond the scope of this study, thus previous surveys were relied on (e.g. Abraham et al. 2009).

5.3.1.3   Budgeting Watershed Investments

The water utility and the inhabitants of Asmara, almost entirely reliant on the Toker Watershed, were assumed to be the most influential, and the most affected stakeholders. Thus, the cost of soil erosion was estimated in terms of the depreciation of asset value. Considering the estimated value of US$44 million of Toker Dam alone, an annual storage capacity loss of 0.5–2% (Abraham et al. 2009) would translate in a reduction of asset value of $220–$880 thousands per year. This is a rough but important estimate, which can be used as a science-informed evidence both for budgeting Watershed Investments and, more strategically, advocating their relevance for the Toker Watershed.

5.3.1.4   Selecting Activities

By way of example, four relevant activities were selected, mainly based on ongoing initiatives in the Toker Watershed. They include (i) *Restoration* through assisted revegetation, where native trees are planted in degraded areas (restoration); (ii) *Protection* of native vegetation, to limit deforestation (protection); (iii) *Terracing*, to reduce erosion in steep areas and (iv) *Agricultural vegetation management*, involving farmers in erosion control measures through voluntary agreements (agricultural vegetation management).

5.3.2   Defining Investment Scenarios

Following the budgetary consideration in the previous section, to cover different stakeholders' level of willingness to invest, six annual budget levels were considered: $10,000; $50,000; $100,000; $250,000; $500,000 and $1,000,000. For budget allocation between activities, two different modalities were applied: *cost-effectiveness* of activities and *pre-allocation* of the entire budget to a single activity at a time. The total number of scenarios can be expressed as $(1 + n) * m * k$, where n, m and k are the number of activities, budget levels and objectives, respectively. In the case study, we considered 60 possible scenarios, given by $(1 + 4) \times 6 \times 2$. At this stage, however, the characterization of the different scenarios based on their feasibility and/or desirability by different stakeholders was not included.

5.4 Biophysical Data Processing and Tailoring of Ecosystem Service Models

In the Asmara and Toker Watershed case study, both the investment objectives are related to soil erosion. Therefore, based on the methods that are implemented in RIOS, the needed input data are LULC, Digital Elevation Model (DEM), rainfall erosivity, soil erodibility, soil depth, USLE C factor and landscape factor (Vogl et al. 2015). Owing to lack of local data, however, some of these input data were obtained from online databases, such as the Harmonized World Soil Database (HWSD), Aster GeDEM (METI and NASA 2011), and studies conducted in other similar contexts (e.g. unit costs of activities adopted from an ongoing Water Fund in Kenya). Table 5.2 summarizes the main biophysical data, specifying resolution and sources, and provides hints on their pre-processing.

RIOS (version 1.1.8) is the tool used for designing *investment portfolios* as well as generating future land use scenarios, whereas InVEST (version 3.2.0) was used to model the impacts on selected ecosystem services. Tables 5.3 and 5.4 present three critical inputs in the RIOS approach, namely, the so-called *transition potentials* and *objective-transition weights* and *activity's unit cost*. The first input defines which activities cause which transitions; the second input specifies the relative contribution of each transition to the objective of the Watershed Investment and the last refers to the overall cost unitary cost of each activity. Similarly, Table 5.5 reports another key input, dealing with additional restriction on watershed activities related, in this specific case, to LULC, slope or elevation.

Indeed, from a boundary work perspective, local knowledge and experience are preferred to better characterize watershed activities. This is particularly valuable for ensuring *saliency* and *legitimacy* during the process of design and assessment of Watershed Investments, and beyond. Owing to lack of local data, however, most inputs used in this case study application were obtained also from online databases, such as the Harmonized World Soil Database (HWSD), and studies conducted elsewhere (e.g. unit costs of activities adopted from an ongoing Water Fund in Kenya).

5.5 Applying Ecosystem Services Models

In the case study application, out of the 60 possible investment scenarios (see Sect. 5.3.2), only 38 scenarios were investigated, and the remaining 22 were found to be unfeasible because of some circumstantial and/or biophysical factors. For instance, areal extension of native vegetation in the Toker Watershed was so small that, a limited budget ($10,000) suffices to cover the whole area. In other cases, increased budget level did not result in a change in the selected ecosystem service.

By way of example, Fig. 5.3 compares two *investment portfolios* aiming at urban water security and rural poverty alleviation, respectively. Both refer to the tenth year, with an annual budget of US$100,000 and budget allocation based on

Table 5.2 Biophysical data for ecosystem services modelling

Data	Source	References	Note
LULC	Africover	FAO GEONETWORK. Spatially Aggregated Multipurpose Landcover Database for Eritrea—AFRICOVER (GeoLayer). (Latest update: 18 Feb 2014) Accessed (21 May 2015). http://data.fao.org/ref/7d456921-5365-4958-8482-799de81dc8af.html?version=1.0	Africover land use and land cover was reclassified using the General LULC classification proposed in RIOS Approach (Vogl et al. 2015)
DEM	Aster Gedem	METI and NASA (2011) ASTER GDEM Version 2. Accessed (2014). http://gdem.ersdac.jspacesystems.or.jp	Hydro-geomorphic analysis using "uDig Spatial Tool box" described in (Abera et al. 2014)
Rainfall erositity	–	Vrieling et al. 2010, 2014	Obtained from (Vrieling et al. 2014, 2010) based on 3-hourly TRMM Multi-satellite Precipitation Analysis precipitation data
Soil erodibility and Soil depth	HWSD	FAO, IIASA, ISRIC, ISSCAS, and JRC. (2012). Harmonized World Soil Database (version 1.2). FAO, Rome, Italy and IIASA, Laxenburg, Austria	Obtained following the analysis described in http://forums.naturalcapitalproject.org/index.php?p= / discussion/comment/ 1384/ #Comment_1384; (last accessed 21/05/2015)

Source Adem Esmail and Geneletti (2017)

Table 5.3 Defining activity's "transition potential" and "unit cost"—Input (1) for "RIOS Investment Portfolio Advisor"

Watershed activity	Transition	Unit cost
Agricultural vegetation management	D—Agricultural vegetation management	US$/ha 125
Protection	A—Keep native vegetation	US$/ha 125
Restoration assisted	C—Revegetation (assisted)	US$/ha 1010
Terracing	E—Ditching	US$/ha 310

Table 5.4 Defining activity's "objective—transition weight"—Input (2) for "RIOS Investment Portfolio Advisor"

Watershed activity	Transitions						
	A	B	C	D	E	F	G
Agricultural vegetation management				1			
Protection	1						
Restoration assisted			1				
Terracing					1		

A—Keep Native Vegetation; B—Revegetation (Unassisted); C—Revegetation (Assisted);
D—Agricultural Vegetation Management; E—Ditching; F—Fertilizer Management;
G—Pasture Management

Table 5.5 Defining land use—and slope-based constraints on activities—Input (3) for "RIOS Investment Portfolio Advisor"

LULC	AG-MGMT	Protection	Restoration	Terracing[§]
Tropical mixed agriculture	Yes	–	Yes	Yes
Permanent crops	Yes	–	–	–
Mixed forest, agriculture	Yes	–	Yes	–
Tropical evergreen forest	–	Yes	Yes	–
Open water	–	–	–	–
Mixed urban	–	–	–	–
Floodplain forest	–	Yes	Yes	–

[§]No terracing for slope less than 12%

cost-effectiveness, and colours representing different watershed activities. Similarly, Fig. 5.4 compares the yearly progress of the same investment portfolios. Both portfolios invest in only two types of activities (i.e. agricultural vegetation management and protection), which also are the least expensive ones (both US$125 per ha against US$310 per ha for terracing or US$1010 per ha for restoration). Spatially, in both cases watershed activities tend to concentrate along the river networks, which is coherent with a generally higher cost-effectiveness of investments in riparian buffers (See model description in Vogl et al. 2015). In the case of rural poverty alleviation, however, there is a marked preference of activities in the two priority sub-watersheds (i.e. *AdiSheka* and *AdiNfas_D01*), which are almost entirely covered, by the end of the 10-year planning period.

Again, Fig. 5.5 shows another illustrative result, allowing exploring the role of the overall budget in shaping the investments portfolio. The example refers to scenarios in which the budget is entirely pre-allocated to assisted restoration. Noteworthy, the marked spatial mismatch between investment portfolios aiming at the two investment objectives. Finally, Tables 5.6 and 5.7 synthesize all the 38 investment portfolios in the case study application, 19 for each Watershed Investment objective. The table

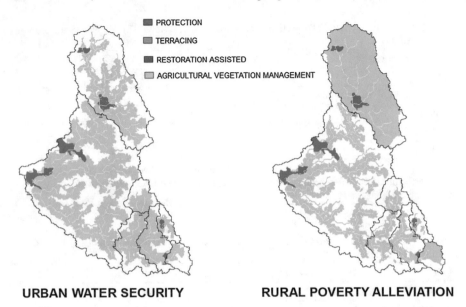

Fig. 5.3 Two illustrative investment portfolios aiming at urban water security (left) and rural poverty alleviation (right), with an annual budget of $100,000, allocated cost-effectively

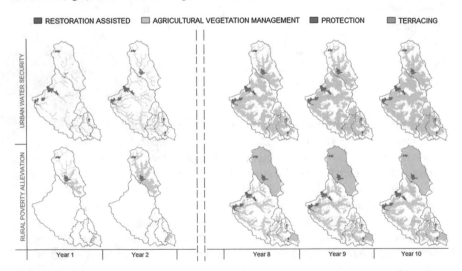

Fig. 5.4 Comparing the yearly progress of two illustrative investment portfolios aiming at urban water security (upper panel) and rural poverty alleviation (lower panel), with an annual budget of $100,000, allocated cost-effectively (*Source* Adem Esmail and Geneletti 2017)

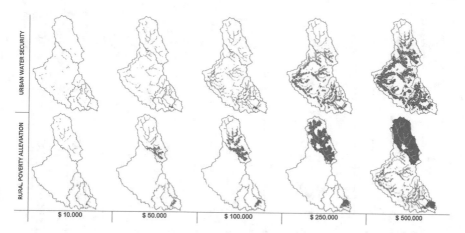

Fig. 5.5 Comparing ten investment portfolios, aiming at the two investment objectives, with five different budget levels, each pre-allocated entirely to restoration (*Source* Adem Esmail and Geneletti 2017)

specifies the budget allocated to each activity as well as the areal extension covered by the activity.

5.6 Assessing the Performance of Investment Scenarios and Follow-Up Planning

A relevant ecosystem service in the case study is soil erosion reduction, which was thus selected as an indicator of performance. In particular, soil erosion reduction corresponds to an enhancement of the ecosystem service, due to a combined effect of a decrease in sediment export and an increase of retention in the watershed (See model description in Vogl et al. 2015). Accordingly, each investment scenario was characterized by the percentage reduction of soil erosion with respect to the baseline conditions, also at sub-watershed level. Table 5.8 summarizes the overall performance of the 38 *investment portfolios*, and, by way of example, the performance at sub-watershed level of the cost-effectiveness scenarios, as well. It highlights the effects of the budget level and budget allocation modality on performance with respect to the two Watershed Investment objectives. For example, in the case of investment for urban water security and an overall budget of $100,000, the reduction of soil erosion is of 15.3% (cost-effectiveness), 19.7% (agricultural vegetation management), 6.6% (assisted restoration) and 9% (terracing). Such values reflect the combined role of the unit cost and of the effectiveness of the activities. Table 5.8 also allows comparing investments aiming at the two objectives, for example, by calculating the difference in performance. A similar analysis has shown that, at watershed level, especially for higher budget levels, the difference tends to be null, and in any case less than 4%.

Table 5.6 Synthesis of investment portfolios for urban water security (activity, allocated budget and areal extension)

Budget allocation[a]	Activity	A		B		C		D		E	
		Budget	Area (ha)	Budget	Area (ha)	Budget	Area (ha)	Budget	Area (ha)	Budget	Area (ha)
$100,000	Ag-mgmt	$56,846	455	$99,900	799	–	–	–	–	–	–
	Protection	$43,054	344	–	–	$57,881	463	–	–	–	–
	Rest-asst	–	–	–	–	–	–	$99,990	99	–	–
	Terracing	–	–	–	–	–	–	–	–	$99,882	322
	Total	$99,900	799	$99,900	799	$57,881	463	$99,990	99	$99,882	322
$500,0000	Ag-mgmt	$442,564	3,541	$499,950	4,000	–	–	–	–	–	–
	Protection	$57,386	459	–	–	–	–	–	–	–	–
	Rest-asst	–	–	–	–	–	–	$499,950	495	–	–
	Terracing	–	–	–	–	–	–	–	–	$499,968	1,613
	Total	$499,950	4,000	$499,950	4,000	–	–	$499,950	495	$499,968	1,613
$1,000,000	Ag-mgmt	$942,244	7,538	$999,900	7,999	–	–	–	–	–	–
	Protection	$57,656	461	–	–	–	–	–	–	–	–
	Rest-asst	–	–	–	–	–	–	$999,900	990	–	–
	Terracing	–	–	–	–	–	–	–	–	$999,936	3,226
	Total	$999,900	7,999	$999,900	7,999	–	–	$999,900	990	$999,936	3,226
$2.500,0000	Ag-mgmt	$1,523,903	12,191	$1,523,903	12,191	–	–	–	–	–	–
	Protection	$57,814	463	–	–	–	–	–	–	–	–
	Rest-asst	–	–	–	–	–	–	$2,499,750	2,475	–	–
	Terracing	–	–	–	–	–	–	–	–	$1,885,956	6,084
	Total	$1,581,716	12,654	$1,523,903	12,191	–	–	$2,499,750	2,475	$1,885,956	6,084

(continued)

Table 5.6 (continued)

Budget allocation[a]		A	B	C	D	E
$5,000,000	Ag-mgmt	–	–	–	–	–
	Protection	–	–	–	–	–
	Rest-asst	–	–	$4,999,500	4,950	–
	Terracing	–	–	–	–	–
	Total	–	–	$4,999,500	4,950	–
$10,000,000	Ag-mgmt	–	–	–	–	–
	Protection	–	–	–	–	–
	Rest-asst	–	–	$9,999,909	9,901	–
	Terracing	–	–	–	–	–
	Total			$9,999,909	9,901	

[a]Budget allocation mode: A. Cost-effectiveness; B. Agricultural vegetation management; C. Protection; D. Terracing; E. Restoration assisted

Table 5.7 Synthesis of investment portfolios for rural poverty alleviation (activity, allocated budget and areal extension)

Budget allocation[a]	Activity	A Budget	A Area (ha)	B Budget	B Area (ha)	C Budget	C Area (ha)	D Budget	D Area (ha)	E Budget	E Area (ha)
$100,000	Ag-mgmt	$83,711	670	$99,900	799	–	–	–	–	–	–
	Protection	$16,189	130	–	–	$57,881	463	–	–	–	–
	Rest-asst	–	–	–	–	–	–	$99,990	99	–	–
	Terracing	–	–	–	–	–	–	–	–	$99,882	322
	Total	$99,900	799	$99,900	799	$57,881	463	$99,990	99	$99,882	322
$500,000	Ag-mgmt	$453,746	3,630	$499,950	4,000	–	–	–	–	–	–
	Protection	$46,204	370	–	–	–	–	–	–	–	–
	Rest-asst	–	–	–	–	–	–	$499,950	495	–	–
	Terracing	–	–	–	–	–	–	–	–	$499,968	1,613
	Total	$499,950	4,000	$499,950	4,000	–	–	$499,950	495	$499,968	1,613
$1,000,000	Ag-mgmt	$942,278	7,538	$999,900	7,999	–	–	–	–	–	–
	Protection	$57,623	461	–	–	–	–	–	–	–	–
	Rest-asst	–	–	–	–	–	–	$999,900	990	–	–
	Terracing	–	–	–	–	–	–	–	–	$999,936	3,226
	Total	$999,900	7,999	$999,900	7,999	–	–	$999,900	990	$999,936	3,226
$2,500,0000	Ag-mgmt	$1,523,903	12,191	$1,523,903	12,191	–	–	–	–	–	–
	Protection	$57,780	462	–	–	–	–	–	–	–	–
	Rest-asst	$545	1	–	–	–	–	$2,499,750	2,475	–	–
	Terracing	–	–	–	–	–	–	–	–	$1,885,956	6,084
	Total	$1,582,228	12,654	$1,523,903	12,191	–	–	$2,499,750	2,475	$1,885,956	6,084

(continued)

Table 5.7 (continued)

Budget allocation[a]		A	B	C	D	E
$5,000,000	Ag-mgmt	–	–	–	–	–
	Protection	–	–	–	–	–
	Rest-asst	–	–	$4,999,500	4950	–
	Terracing	–	–	–	–	–
	Total	–	–	$4,999,500	4,950	–
$10,000,000	Ag-mgmt	–	–	–	–	–
	Protection	–	–	–	–	–
	Rest-asst	–	–	$9,999,909	9901	–
	Terracing	–	–	–	–	–
	Total			$9,999,909	9,9001	

[a]Budget allocation mode: A. Cost-effectiveness; B. Agricultural vegetation management; C. Protection; D. Terracing; E. Restoration assisted

Table 5.8 Synthesis of performance of the proposed Watershed Investments: percentage reduction of soil erosion at sub-watershed level

	Sub-watershed	Urban water Ecurity						Rural poverty alleviation					
		$10 (%)	$50	$100	$250	$500	$1000	$10 (%)	$50	$100	$250	$500	$1000
Cost-effectiveness	Toker	−0.7	2.6	−10.1%	−24.5%	–	–	3.3	−6.1%	−15.8%	−24.5%	–	–
	AdiSheka	−5.5	−29.0%	−29.6%	−32.2%	–	–	−0.3	−4.0%	−17.4%	−32.2%	–	–
	Maisirwa	−2.8	−8.7%	−14.1%	−28.6%	–	–	−0.3	−5.3%	−14.0%	−28.6%	–	–
	Beleza	−2.3	−7.6%	−18.9%	−33.8%	–	–	0.4	1.6%	1.6%	0.9%	–	–
	Valle-Gnocchi	0.0	−0.4%	1.2%	0.9%	–	–	−1.6	−10.0%	−24.3%	−33.8%	–	–
	AdiNifas_Do1	0.4	0.4%	0.4%	0.4%	–	–	0.4	0.4%	0.4%	0.4%	–	–
	AdiNifas_D02	0.0	4.9%	5.0%	5.0%	–	–	5.0	5.0%	5.0%	5.0%	–	–
	Total	−2.2	−7.3%	−15.3%	−25.4%	–	–	1.8	−5.0%	−15.2%	−25.4%	–	–
Agri. Veg. management	Toker	−0.8	−2.6%	−16.6%	−29.6%	–	–	−2.1	−12.4%	−21.8%	−29.6%	–	–
	AdiSheka	−8.3	−30.5%	−31.1%	−33.6%	–	–	−1.6	−6.8%	−20.3%	−33.6%	–	–
	Maisirwa	−2.3	−8.9%	−15.0%	−28.6%	–	–	−0.7	−6.2%	−15.5%	−28.6%	–	–
	Beleza	−2.7	−8.7%	−20.5%	−33.8%	–	–	0.0	0.0%	0.0%	−0.7%	–	–
	Valle-Gnocchi	0.0	−0.4%	−0.4%	−0.7%	–	–	−2.5%	−11.9%	−25.8%	−33.8%	–	–
	AdiNifas_Do1	0.0	0.0%	0.0%	0.0%	–	–	0.0	0.0%	0.0%	0.0%	–	–
	AdiNifas_D02	0.0	0.0%	0.0%	0.0%	–	–	0.0	0.0%	0.0%	0.0%	–	–
	Total	−3.0	−10.9%	−19.7%	−28.8%	–	–	−1.7	−9.7%	−19.7%	−28.8%	–	–
Protection	Toker	5.0	–	–	–	–	–	5.0	–	–	–	–	–
	AdiSheka	1.4	–	–	–	–	–	1.4	–	–	–	–	–
	Maisirwa	0.0	–	–	–	–	–	0.0	–	–	–	–	–
	Beleza	0.0	–	–	–	–	–	1.6	–	–	–	–	–

(continued)

Table 5.8 (continued)

	Sub-watershed	Urban water Ecurity						Rural poverty alleviation					
		$10 (%)	$50	$100	$250	$500	$1000	$10 (%)	$50	$100	$250	$500	$1000
	Valle-Gnocchi	1.6	–	–	–	–	–	0.0	–	–	–	–	–
	AdiNifas_Do1	0.4	–	–	–	–	–	0.4	–	–	–	–	–
	AdiNifas_D02	5.0	–	–	–	–	–	5.0	–	–	–	–	–
	Total	3.4	–	–	–	–	–	3.4	–	–	–	–	–
Assisted restoration	Toker	−0.1	−53.5%	−58.9%	−69.8%	−75.7%	−36.0%	−0.4	−1.9%	−4.1%	−12.2%	−23.8%	−41.7%
	AdiSheka	−1.4	−89.8%	−90.2%	−90.8%	−91.6%	−53.8%	−0.2	−2.2%	−2.8%	−5.6%	−18.0%	−44.9%
	Maisirwa	−0.2	1205.8%	1197.0%	1190.5%	1104.2%	−38.6%	−0.1	−0.6%	−2.1%	−6.9%	−17.8%	−39.3%
	Beleza	−0.6	31.5%	31.5%	31.3%	31.7%	−0.6%	0.0	0.0%	0.0%	0.0%	0.0%	−0.1%
	Valle-Gnocchi	0.0	−54.6%	−54.6%	−54.6%	−54.6%	−48.1%	−1.1	−3.1%	−5.5%	−12.1%	−26.3%	−52.0%
	AdiNifas_Do1	0.0	0.0%	0.0%	0.0%	0.0%	0.0%	0.0	0.0%	0.0%	0.0%	0.0%	0.0%
	AdiNifas_D02	0.0	132.8%	123.2%	105.7%	87.1%	0.0%	0.0	0.0%	0.0%	0.0%	0.0%	0.0%
	Total	−0.5	−2.9%	−6.6%	−13.5%	−21.4%	−39.1%	−0.3	−1.8%	−3.4%	−9.2%	−20.3%	−40.0%
Terracing	Toker	−0.4	−57.9%	−59.1%	−60.3%	–	–	−0.9	−5.4%	−9.3%	−14.9%	–	–
	AdiSheka	−3.5	−90.1%	−90.3%	−91.1%	–	–	−1.0	−3.0%	−9.7%	−20.4%	–	–
	Maisirwa	−1.4	1208.6%	1145.7%	1022.4%	–	–	−0.2	−3.3%	−6.7%	−17.2%	–	–
	Beleza	−1.5	31.4%	31.1%	30.8%	–	–	0.0	0.0%	0.0%	−0.5%	–	–
	Valle-Gnocchi	0.0	−54.6%	−54.6%	−54.5%	–	–	−0.8	−3.5%	−8.8%	−14.7%	–	–
	AdiNifas_Do1	0.0	0.0%	0.0%	0.0%	–	–	0.0	0.0%	0.0%	0.0%	–	–
	AdiNifas_D02	0.0	130.7%	124.7%	101.2%	–	–	0.0	0.0%	0.0%	0.0%	–	–
	Total	−1.3	−5.4%	−9.0%	−15.5%	–	–	−0.8	−4.2%	−8.7%	−15.5%	–	–

This is an important piece of information, which would need further investigation. A possible explanation could be that at higher budget levels such large portions of the watershed were covered for both objectives, while at lower budgets, the overall impact of the investments was minimum.

However, when making any comparison based on Table 5.8, it is worth bearing in mind that, for a given budget level, the money actually invested could differ based on the scenario. Again, it is should be recalled that all the analysis were based on uncalibrated models and assumptions regarding the unit costs of activities. An area for further stakeholder involvement could include observed data for calibration and validation of results and data on activities' cost and effectiveness.

At this stage, the aggregated performance values in Table 5.8 combined with the feasibility and/or desirability considerations of stakeholders are essential to reach the necessary consensus over the actual Watershed Investment to be implemented. Figure 5.6 represents an overview of the performance of the 38 investments scenarios, at a sub-watershed level. We here recall that the comparison between different scenarios is also supported by spatially and temporally explicit data, such as maps of activities, land use scenarios and impacts on the selected ecosystem services. Such data provide detailed guidance for the actual implementation of the Watershed Investments and at the same time form a strong basis for planning a follow-up. Interestingly, the outputs of the technical component can easily feed the follow-up planning process, typically consisting of four elements (Morrison-Saunders et al. 2007), namely, *monitoring* (i.e. collection of pre and post-implementation activity

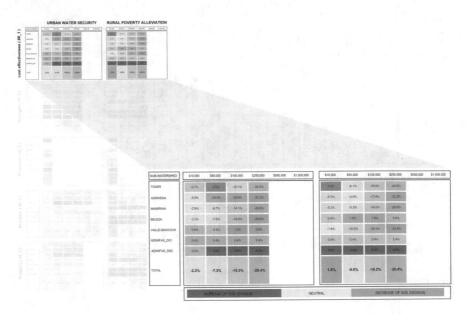

Fig. 5.6 Overview of the analysed investment scenario and an example of a comparison of different Investment scenarios at sub-watershed level

and environmental data); *evaluation* (i.e. appraisal of the conformance with standards, predictions or expectations); *management* (i.e. making decisions and taking appropriate action in response to issues arising from monitoring and evaluation) and *communication* (i.e. updating the stakeholders to provide feedback on project and process).

5.7 Lessons Learned and Conclusions

From an ecosystem services perspective, this chapter makes an important contribution by illustrating a case study application that addresses some of the main hindrances to the operationalization of the concept. According to Turner and Daily (2008), for example, three key hindrances are represented by an *information failure*, i.e. lack of detailed information at scales relevant to decision-making; a *market failure*, i.e. lack of compelling models of success, aligning economic incentives with conservation; and an *institutional failure*, lack of practical know-how in the process of institutional design and implementation. The case study application directly addressed the *information failure*, providing detailed answers to some key management and planning questions; at the same time, it contributed to tackling the other two failures, as well. The simple approach for budgeting Watershed Investments that accounts for the loss aversion bias of stakeholders, recalling the New York City case study where budgeting was favourably influenced by an ecosystem services approach, is a good example.

In addition, this chapter illustrated practical application in the urban water sector of the concept of boundary work proposed by Clark et al. (2016), which provides guidance on collaborative knowledge generation for an adaptive ecosystem management, and establishment of good working relations among diverse stakeholders to promote cooperative implementation. By way of example, we mention here the distinction between two components (strategic and technical) and three stages, which reflect the different needs of boundary work, in order to effectively and timely facilitate negotiation among stakeholders engaged in knowledge use and production. The strategic component mainly ensures *saliency* and *legitimacy*, while the technical component ensures *credibility*. Yet, the two components are tightly interlinked. Actually, only jointly do they contribute to a successful implementation of Watershed Investments, by linking diverse sets of stakeholders and knowledge systems, across different management levels and institutional boundaries (Kowalski and Jenkins 2015). As Parker and Crona (2012) put it, boundary work is a dynamic process that takes place in a *landscape of tensions*, rather than a single-time achievement. Thus, the importance of gaining a good understanding of the contextual and contingent factors of the specific socio-ecological system, as well as the relative influence of the involved social actors, to determine the different boundary work needs, as was done in Chap. 4.

More specifically, the empirical testing of the technical component illustrated the applicability of the proposed approach to medium-sized cities in a data-poor context in Sub-Saharan Africa. As mentioned in the introduction in Chap. 1, these are the cities that are expected to host the majority of urban population growth in the next decade (United Nations, DESA, Population Division 2018). We consciously selected the case study knowing that data paucity and resource scarcity are two challenging factors. Again, soil erosion- and water scarcity-related issues are among the most common issues affecting many medium-sized cities in Africa. The case study application highlighted the main challenges in terms of data paucity and boundary work that should be put in place to facilitate the negotiation among stakeholders. In particular, data paucity affected the results of the analysis, for example, some data used in the analysis had low resolution such as the case of the rainfall erosivity provided by Vrieling et al. (2010). Boundary workwise, local knowledge and experience should be preferred to better characterize watersheds and watershed activities. This is highly valuable for ensuring *saliency* and *legitimacy* during the process of design and assessment of Watershed Investments, and beyond. Noteworthy, our analysis did not include calibration of the models and relied on some assumptions regarding, for instance, the unit costs of activities. Accordingly, another area for further stakeholder input could involve obtaining observed data for calibration and validation of results, as well as data on cost of activities and their effectiveness.

Despite these limitations, the application highlighted the potential of the proposed approach to support implementation of Watershed Investments, and ultimately promote adaptive watershed management. In this regard, in Chap. 4, three illustrative watershed initiatives that could represent windows of opportunity for applying the proposed approach in the Toker Watershed were identified. These three watershed initiatives are representative of the contextual and contingent factors as well as the relative influence of stakeholders in the case study. A first initiative consists of existing partnerships between the Asmara water utility and farmers in the Central Region, in which the Toker Watershed is located. A second initiative is a transdisciplinary research project dealing with water resource management in the *Upper Anseba Watershed* of which the Toker is a sub-watershed. Finally, a third initiative is the so-called Summer Student Work Program (SSWP).

From an adaptive management perspective, these three initiatives provide an interesting point of departure. As previously indicated in Chap. 4, singularly, the three watershed initiatives could benefit from the application of the here proposed operational approach for designing and assessing impacts of Watershed Investments. Possibly, this ecosystem services- and boundary work-based approach, could trigger a process of social learning, involving the stakeholders in these three initiatives, within a framework of adaptive watershed management. Eventually, the AWSD would be a central actor. Beyond the specific case, we hope that this study will inspire other medium-sized cities in Sub-Saharan Africa to experiment innovative and creative ways of facing their challenges, based on the concepts of ecosystem services and boundary work.

References

Abera W, Antonello A, Franceschi S, Formetta G, Rigon R (2014) The uDig spatial toolbox for hydro-geomorphic analysis. Geomorphol Tech 1

Abraham D, Tesfaslasie F, Tesfay S (2009) An appraisal of the current status and potential of surface water in the upper Anseba catchment potential of surface water in the upper Anseba catchment. https://doi.org/10.7892/boris.69703

Adem Esmail B, Geneletti D (2017) Design and impact assessment of watershed investments: an approach based on ecosystem services and boundary work. Environ Impact Assess Rev 62:1–13. https://doi.org/10.1016/j.eiar.2016.08.001

Clark WC, Tomich TP, van Noordwijk M, Guston D, Catacutan D, Dickson NM, McNie E (2016) Boundary work for sustainable development: natural resource management at the Consultative Group on International Agricultural Research (CGIAR). Proc Natl Acad Sci 113:4615–4622. https://doi.org/10.1073/pnas.0900231108

IPCC (2014) Summary for policy makers. In: Field CB, Barros VR, Dokken DJ, KJ, M, Mastrandrea MD, Bilir TE, Chatterjee M, Ebi KL, Estrada YO, Genova RC, Girma B, Kissel ES, Levy AN, MacCracken S, Mastrandrea PR, White LL (eds) Climate change 2014: impacts, adaptation and vulnerability—contributions of the working group II to the fifth assessment report of the intergovernmental panel on climate change. Cambridge University Press, Cambridge, UK, New York, NY, USA, pp 1–32

Kowalski AA, Jenkins LD (2015) The role of bridging organizations in environmental management: examining social networks in working groups. Ecol Soc 20:16. https://doi.org/10.5751/ES-07541-200216

METI and NASA (2011) ASTER GDEM Version 2. Accessed (2014). http://gdem.ersdac.jspacesystems.or.jp

MoLWE (2012) Eritrea's second national communication under the United Nations framework convention on climate change (UNFCCC). Asmara, Eritrea

Morrison-Saunders A, Marshall R, Arts J (2007) EIA follow-up international best practice principles. Special Publication Series No. 6. Int Assoc Impact Assess

Murtaza N (1998) The pillage of sustainablility in Eritrea, 1600s–1990s: rural communities and the creeping shadows of hegemony. Greenwood Press, Westport, Connecticut

Parker J, Crona B (2012) On being all things to all people: boundary organizations and the contemporary research university. Soc Stud Sci 42:262–289. https://doi.org/10.1177/0306312711435833

Tewolde MG, Cabral P (2011) Urban sprawl analysis and modeling in asmara. Eritrea Remote Sens 3:2148–2165. https://doi.org/10.3390/rs3102148

Turner RK, Daily GC (2008) The ecosystem services framework and natural capital conservation. Environ Resour Econ 39:25–35. https://doi.org/10.1007/s10640-007-9176-6

United Nations, Department of Economic and Social Affairs, Population Division (2018) World Urbanization Prospects: The 2018 Revision, Online Edition

Vogl A, Tallis H, Douglass J, Sharp R, Wolny S, Veiga F, Benitez S, León J, Game E, Petry P, Guimerães J, Lozano JS (2015) Resource investment optimization system (RIOS). In: v1.1.0. Introduction and theoretical documentation. Stanford, CA

Vrieling A, Sterk G, de Jong SM (2010) Satellite-based estimation of rainfall erosivity for Africa. J Hydrol 395:235–241. https://doi.org/10.1016/j.jhydrol.2010.10.035

Vrieling A, Hoedjes JCB, van der Velde M (2014) Towards large-scale monitoring of soil erosion in Africa: accounting for the dynamics of rainfall erosivity. Glob Planet Change 115:33–43. https://doi.org/10.1016/j.gloplacha.2014.01.009

Chapter 6
Conclusions

Abstract This chapter summarizes the main messages of the book, as well as discusses the challenges for future research and practice to contribute to achieving water security and to implementing adaptive management in the urban water sector. Briefly, the first main message is that achieving urban water security through adaptive watershed planning and management, in Sub-Saharan Africa context, is a complex issue. Thus, an intuitive and flexible conceptual framework of the urban water sector from an ecosystem services perspective was proposed. It provides an overview of the main challenges and trends that characterize the sector, highlighting the specificities of the Sub-Saharan context, setting the background for further analysis. Second, if properly designed, Watershed Investments can become an important financial and governance mechanism to promote the implementation of adaptive watershed management to achieve urban water security. Third, a good case study application, even if only based on desk research, can serve to inspire stakeholder and possibly prepare the ground for real-life implementation of science-informed measures to promote urban water security alongside other social goals, coordinating ongoing watershed initiatives.

Keywords Urbanization · Ecosystem services · Watershed investments · Water security · Sub-Saharan Africa

The ultimate goal of this book was to contribute to addressing the urgent challenges of urban water security facing many medium-sized cities in Sub-Saharan Africa. The book focused on Watershed Investments to secure water for cities, which consist of governance and financial mechanisms that secure clean water for cities, and operate by engaging upstream communities. Importantly, Watershed Investments are acknowledged as a promising opportunity to effect large-scale transformative changes for sustainable development. Accordingly, this book pursued three specific objectives, namely, (i) developing a conceptual framework of the urban water sector from an ecosystem services perspective, highlighting the specificities of the Sub-Saharan context; (ii) developing an operational approach for designing and assessing impacts of Watershed Investments, based on ecosystem services and boundary work and (iii) testing the proposed approach through a case study of the urban sector form Sub-Saharan Africa.

Towards the achievement of these objectives, the book explored and jointly applied three novel concepts of ecosystem services, boundary work and learning organizations. The concept of ecosystem services provided a holistic approach for framing socio-ecological issues and for integrating different types of data (e.g. biophysical and socio-economic), to identify possible solutions, for example, in the form of different portfolios of Watershed Investments. Boundary work, as active management of tension arising at the interface between stakeholders to facilitate transfer of knowledge into action, served to develop an operational approach for a participatory and iterative process of designing and assessing impacts of Watershed Investments. Finally, learning organization, and related concept of institutional capacity, helped to frame the role of water utilities, as social actors, in structuring their choice of action within a society to pursue their mission in terms of urban water security.

In the remainder of this chapter, the main findings grouped by the three specific objectives are summarized and some recommendations are discussed for future research.

6.1   A Conceptual Framework of the Urban Water Sector

Achieving urban water security through adaptive watershed planning and management is a complex issue. Thus, an intuitive and flexible conceptual framework of the urban water sector from an ecosystem services perspective was built. It provides an overview of the main challenges and trends that characterize the sector, highlighting the specificities of the Sub-Saharan context, thus setting the background for further analysis.

The first part of the book aimed at gaining an overview of the urban water sector and the complexities it entails. Thus, it built an intuitive and flexible conceptual framework of the urban water sector from an ecosystem services perspective, based on an original review of the literature. The proposed framework attempted to synthesize the most relevant aspects characterizing the exchange of water between watersheds and cities, and within the city. It highlighted the role of urban water infrastructures in (i) linking ecosystem services production and benefit areas, (ii) in bridging spatial scales ranging from the watershed to the household level and (iii) in adopting ecosystem-based responses to water vulnerability. Noteworthy, is it built on internationally accepted frameworks (e.g. SEEA-Water) and concepts (e.g. Integrated Urban Water Management) and, at the same time, it took the ease of application into account (e.g. use of layperson terms). The framework attempted to be as simple, intuitive and flexible as possible therefore it has a good potential to be used as a tool for involving stakeholders. Finally, an illustrative application as a tool for reviewing real-life infrastructural projects showed the potential and the limits of the framework.

Attempting to represent an entire "management paradigm", the findings of this part of the research and the proposed framework are arguable, not exhaustive and provide an overall idea of the urban water sector. Nevertheless, despite these limitations, the findings and the proposed framework can be a useful starting point for seeking a better understanding of the complex relationship between long-term human wellbeing in cities and the respective service providing and life-supporting watersheds.

6.2 An Operational Approach for Designing Watershed Investments

If properly designed, Watershed Investment can become an important financial and governance mechanism to promote the implementation of adaptive watershed management to achieve urban water security.

In Chap. 3, an operative approach for designing and assessing the impacts of Watershed Investments was built. The proposed process-based approach builds on spatially explicit modelling of ecosystem services and insights on boundary work. It is structured to facilitate negotiations among stakeholders: distinguishing between a strategic component addressing concerns of *saliency* and *legitimacy* and a technical component ensuring *credibility*, respectively. The former includes setting the agenda, defining investment scenarios and assessing the performance of Watershed Investments. The latter concerns data processing and preparation, tailoring spatially explicit ecosystem services models, hence applying them to design a set of *investment portfolios*, generate future land use scenarios and model impacts on selected ecosystem services.

The proposed approach, by addressing stakeholders' concerns of *credibility*, *saliency* and *legitimacy*, is expected to facilitate negotiation of objectives, definition of scenarios and assessment of alternative Watershed Investments. Ultimately, it can contribute to implementing adaptive watershed management thorough Watershed Investments, which are considered a promising opportunity for achieving large-scale transitions towards sustainability, in the near future.

6.3 A Case Study Application to Inspire Real-Life Implementation

A good case study application can serve to inspire stakeholder and possibly prepare the ground for real-life implementation of science-informed measures to promote urban water security alongside other social goals, coordinating ongoing watershed activities.

In Chaps. 4 and 5, the aim was to test the proposed approach through a case study of the urban sector form Sub-Saharan Africa. A case study involving a medium-sized city and its watershed in a data-scarce context in Sub-Saharan Africa was selected:

Asmara city and the Toker watershed in Eritrea. Specifically, soil erosion and water scarcity-related challenges were associated to two illustrative Watershed Investment objectives: urban water security and rural poverty alleviation. Beyond the specific case study, Chap. 4 also contributed to gaining a better understanding of the role of water utilities as learning organization. Thus, the Water Utility Model (WUM) was introduced and applied as tool for evaluating institutional capacity of the Asmara Water Supply Department, as a central actor that could potentially drive the implementation of an adaptive management in the case study.

As illustrated in Chap. 5, the case study application produced spatially explicit data: investment portfolio, land use scenario and impact on ecosystem services. Such data were aggregated to quantitatively assess the performance of Watershed Investments, in terms of changes in a selected ecosystem service; thus answering key management and planning questions.

Besides the limitation related to a single case study approach, the main weakness of this part of the research is the limited involvement of actual stakeholders. This has led to different assumptions concerning, for example, the selection of illustrative investment objectives, reclassification of land use and the unit cost of activities. Another limit was the coarse resolution of some of the data used for the modelling (e.g. rainfall erosivity), and the fact that the models were not calibrated. Again, the focus on a single ecosystem service, i.e. soil erosion control, seems to be at odd with the holistic approach of the ecosystem services approach. Therefore, the results are only illustrative of the potential application.

On the others hand, the limited degree of generalizability was addressed by focusing on a medium-sized city in a Sub-Saharan Africa context and considering two of the most common socio-ecological challenges affecting such cities: soil erosion and water scarcity. In fact, the case study application highlighted the main challenges in terms of data paucity as well as boundary work that should be put in place to facilitate the negotiation among stakeholders. In particular, building on real-life experiences and tools developed by the Water Funds, mainly in Latin America, it showed a possible operationalization of the ecosystem services approach for watershed management in data-poor contexts in Africa. Further research, involving actual stakeholders in the case study would allow to test the here proposed operative approach.

Printed in the United States
By Bookmasters